T0257797

Encyclopedia of Remote Sensing: Analysis Techniques Volume I

Edited by **Matt Weilberg**

New York

Published by Callisto Reference,
106 Park Avenue, Suite 200,
New York, NY 10016, USA
www.callistoreference.com

Encyclopedia of Remote Sensing: Analysis Techniques
Volume I
Edited by Matt Weilberg

International Standard Book Number: 978-1-63239-288-6 (Hardback)

Contents

Preface

This book aims to highlight the current researches and provides a platform to further the scope of innovations in this area. This book is a product of the combined efforts of many researchers and scientists, after going through thorough studies and analysis from different parts of the world. The objective of this book is to provide the readers with the latest information of the field.

This book gives a comprehensive account of various areas related to remote sensing. It attempts to cover major aspects of remote sensing techniques by providing detailed studies about new techniques for data processing. It includes contributions of renowned researchers, experts and practitioners in this field from all over the world. Their collective expertise impart significant knowledge and serve as a valuable reference for researchers, students and other interested individuals in this field.

I would like to express my sincere thanks to the authors for their dedicated efforts in the completion of this book. I acknowledge the efforts of the publisher for providing constant support. Lastly, I would like to thank my family for their support in all academic endeavors.

Editor

Analysis Techniques

Statistical Properties of
Surface Slopes via Remote Sensing

Josué Álvarez-Borrego[1] and Beatriz Martín-Atienza[2]
[1]CICESE, División de Física Aplicada,
Departamento de Óptica
[2]Facultad de Ciencias Marinas, UABC
México

1. Introduction

The complexity of wave motion in deep waters, which can damage marine platforms and vessels, and in shallow waters, same that can afflict human settlements and recreational areas, has given origin to a long-term development in laboratory and field studies, the conclusions of which are used to design methodology and set bases to understand wave motion behavior.

Via remote sensing, the use of radar images and optical processing of aerial photographs has been used. The interest in wave data is manifold; one element is the inherent interest in the directional spectra of waves and how they influence the marine environment and the coastline. These wave data can be readily and accurately collected by aerial photographs of the wave sun glint patterns which show reflections of the Sun and sky light from the water and thus offer high-contrast wave images.

In a series of articles, Cox and Munk (1954a, 1954b, 1955) studied the distribution of intensity or glitter pattern in aerial photographs of the sea. One of their conclusions was that for constant and moderate wind speed, the probability density function of the slopes is approximately Gaussian. This could be taken as an indication that in certain circumstances, the ocean surface could be modeled as a Gaussian random process. Similar observations by Longuet-Higgins et al. (1963) (cited by Longuet-Higgins (1962)) with a floating buoy, which filters out the high-frequency components, come considerably closer to the Gaussian distribution.

Other authors (Stilwell, 1969; Stilwell & Pilon, 1974) have studied the same problem considering a sea surface illuminated by a continuous sky light with no azimuthal variations in sky radiance. Different models of sky light have been used emphasizing the existence of a nonlinear relationship between the slope spectrum and the corresponding wave image spectrum (Peppers & Ostrem, 1978; Chapman & Irani, 1981).

Simulated sea surfaces have been analyzed by optical systems to understand the optical technique in order to obtain best qualitative information of the spectrum (Álvarez-Borrego, 1987; Álvarez-Borrego & Machado, 1985).

Fuks and Charnotskii (2006) derived the joint probability density function of surface height and partial second derivatives for an ensemble of specular points at a random rough Gaussian isotropic surface at normal incidence. However, in a real physical situation, consideration of Gaussian statistics can be a very good approximation.

Cox and Munk (1956) observed that the center of the glitter pattern images had shifted downwind from the grid center. This shift can be associated with an up/downwind asymmetry of the wave profile (Munk, 2009). Surfaces of small positive slope are more probable than those of negative slope; large positive slopes are less probable than larger negative slopes, thus permitting the restraint of a zero mean slope (Bréon & Henrist, 2006).

According with Longuet-Higgins (1963) the sea surface slopes have a Gaussian probability function to a first approximation. In the next approximation skewness is taken into account. The kurtosis is zero, as are all the higher cumulants. In the next approximation, the distribution is given taken into account the kurtosis.

Walter Munk (2009) writes that the skewness appears to be correlated with a rather sudden onset of breaking for winds above 4 m s[-1] and he does not think that skewness comes from parasitic capillaries. Chapron et al. (2002) suggest that the actual waves form under near-breaking conditions, along with the varying population and length scales for these breaking events, should also contribute to the skewness.

In this chapter we will consider two different cases to analyze statistical properties of surface slopes via remote sensing: first we assume the fluctuation of the surface slopes to be statistically Gaussian and the second case we assume the fluctuation of the surface slopes to be statistically non-Gaussian. We, also, assume that the surfaces are illuminated by a source, the Sun, of a fixed angular extent, β, and imaged through a lens that subtends a very small solid angle. With these considerations, we calculated their images, as they would be formed by a signal clipping detector. In order to do this, we define a "glitter function", which operates on the slope of the surfaces. In the first case we consider two situations: the detector line of sight angle, θ_d, is constant for each point on the surface and θ_d is variable for each point in the surface. In the second case, with non-Gaussian statistics, we consider θ_d variable for each point in the surface only, because we consider that this case is more realistic.

2. Geometry of the model (Gaussian case considering a constant detector angle)

The physical situation is shown in figure 1. The surface $\zeta(x)$ is illuminated by a uniform incoherent source S of limited angular extent, with wavelength $\overline{\lambda}$. Its image is formed in D by an aberration free optical system. The incidence angle, θ_s, is defined as the angle between the incidence angle direction and the normal to the mean surface. Then, in figure 1, θ_s, represents the mean angle subtended by the source S and θ_d represents the mean angle subtended by the optical system of the detector with the normal to the mean surface.

The apparent diameter of the source is β and of the detector is δd. Light from the source is reflected on the surface just one time and, depending on the slope, the light reflected will or will not be part of the image. In broad terms, the image consists of bright and dark regions that we call a glitter pattern. α represents the angle between the x axis and the surface, and

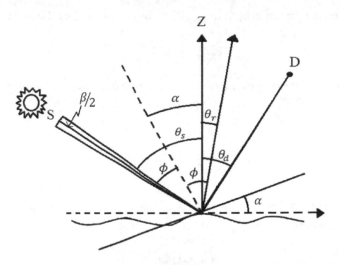

Fig. 1. The detector is located in the zenith of each reflection point in the profile.

ϕ represents the angle between the normal to the plane and the source S. This angle is given by $\phi = \theta_s - \alpha$, and the specular angle is given by $\phi = \theta_r + \alpha$. From this two equations we can write

$$\theta_r = \theta_s - 2\alpha \ .$$
(1)

Because the source has a finite size, there are several incidence directions which are specular reflected to the camera. The directions, θ_{os} (where this angle is the angular dimension of the Sun), where there are incidence rays which are determined by the condition

$$\theta_s - \frac{\beta}{2} \leq \theta_{os} \leq \theta_s + \frac{\beta}{2},$$
(2)

in other words, the source is angularly described by the function, $\sigma(\theta_{os})$, can be written like

$$\sigma(\theta_{os}) = rect\left[\frac{\theta_{os} - \theta_s}{\beta} \right],$$
(3)

where rect(.) represents the rectangle function (Gaskill, 1978).

So, the projection of this source on the detector, after reflection, is given by

$$\theta_s - \frac{\beta}{2} - 2\alpha \leq \theta \leq \theta_s + \frac{\beta}{2} - 2\alpha \ ,$$
(4)

$$\sigma_R(\theta) = rect\left(\frac{\theta - \theta_r}{\beta} \right),$$
(5)

where equation (1) is taken into account.

On the other side, the detection system pupil can be represented by the function

$$P(\theta) = rect\left(\frac{\theta - \theta_d}{\delta d}\right). \tag{6}$$

The intensity light I, arriving to the detection plane D depends on the overlap between the functions $\sigma_R(\theta)$ and $P(\theta)$, and can be approximated by

$$I = \int_{-\frac{\pi}{2}}^{\frac{\pi}{2}} \sigma_R(\theta) P(\theta) d\theta. \tag{7}$$

In practical situations δd is so smaller than β, that we can to approximate $P(\theta) = \delta(\theta - \theta_d)$, where δ is the Dirac delta, of this way

$$I \approx \sigma_R(\theta_d),$$
$$\approx rect\left(\frac{\theta_d - \theta_r}{\beta}\right). \tag{8}$$

The light reflection will arrive to the detector D when

$$\theta_r - \frac{\beta}{2} \le \theta_d \le \theta_r + \frac{\beta}{2}, \tag{9}$$

and because $\theta_r = \theta_s - 2\alpha$, we have

$$\frac{\theta_s - \theta_d}{2} - \frac{\beta}{4} \le \alpha \le \frac{\theta_s - \theta_d}{2} + \frac{\beta}{4}. \tag{10}$$

Defining $\Pi = \tan\alpha$, $\gamma = (\theta_s - \theta_d)/2$ and $\Pi_o = \tan\gamma$, and using the relationship $\tan(\gamma \pm \beta/4) \approx \tan\gamma \pm (1 + \tan^2\gamma)\beta/4$, valid for small $\beta/4$, we obtain the next condition for the slopes

$$\Pi_o - (1 + \Pi_o^2)\frac{\beta}{4} \le \Pi \le \Pi_o + (1 + \Pi_o^2)\frac{\beta}{4}. \tag{11}$$

We find then the "glitter function", given by

$$B(\Pi) = rect\left[\frac{\Pi - \Pi_o}{(1 + \Pi_o^2)\beta/2}\right]. \tag{12}$$

This expression (eq. 12) tell us that the geometry of the problem selects a surface slope region and encodes like bright points in the image (glitter pattern).

2.1 Relationship among the variances of the intensities in the image, surface slopes and surface heights

The mean of the image, μ_I, may be written (Papoulis, 1981)

$$\mu_I = \langle I(x) \rangle = \int_{-\infty}^{+\infty} B(\Pi) p(\Pi) d\Pi, \tag{13}$$

where $B(\Pi)$ is defined by equation (12) and $p(\Pi)$ is the probability density function in one dimension, where in a first approximation a Gaussian function is considered. Substituting in equation (13) the expressions for $B(\Pi)$ and $p(\Pi)$, we have

$$\mu_I = \langle I(x) \rangle = \frac{1}{\sigma_\Pi (2\pi)^{1/2}} \int_{-\infty}^{\infty} rect \left[\frac{\Pi - \Pi_o}{\left(1 + \Pi_o^2\right)^{\beta/2}} \right] \exp\left(-\frac{\Pi^2}{2\sigma_\Pi^2} \right) d\Pi. \tag{14}$$

Defining $a = \Pi_o - \left(1 + \Pi_o^2\right)(\beta/4)$ and $b = \Pi_o + \left(1 + \Pi_o^2\right)(\beta/4)$, we can write

$$\mu_I = \langle I(x) \rangle = \frac{1}{2} \left[erf\left(\frac{b}{\sqrt{2}\sigma_\Pi} \right) - erf\left(\frac{a}{\sqrt{2}\sigma_\Pi} \right) \right]. \tag{15}$$

The variance of the intensities in the image, σ_I^2, is defined by (Papoulis, 1981)

$$\sigma_I^2 = \langle I^2(x) \rangle - \langle I(x) \rangle^2 = \int_{-\infty}^{+\infty} \left[B(\Pi) - \mu_I \right]^2 p(\Pi) d\Pi. \tag{16}$$

But, $B(\Pi) = B^2(\Pi)$, then $\langle I^2(x) \rangle = \langle I(x) \rangle$, therefore

$$\sigma_I^2 = \langle I(x) \rangle - \langle I(x) \rangle^2 = \mu_I (1 - \mu_I), \tag{17}$$

and substituting the expression of $\langle I(x) \rangle$, equation (15), in equation (17), we have

$$\sigma_I^2 = \frac{1}{2} \left[erf\left(\frac{b}{\sqrt{2}\sigma_\Pi} \right) - erf\left(\frac{a}{\sqrt{2}\sigma_\Pi} \right) \right] - \left[\frac{1}{2} \left[erf\left(\frac{b}{\sqrt{2}\sigma_\Pi} \right) - erf\left(\frac{a}{\sqrt{2}\sigma_\Pi} \right) \right] \right]^2, \tag{18}$$

which is the required relation between the variance of the intensities in the image, σ_I^2, and the variance of the surface slopes, σ_Π^2.

The relation (18) is shown in figure 2 for some typical cases, using the geometry described above, with $\theta_d = 0^o$ and $\beta = 0.68^o$. In the horizontal axis we have the variance of the surface slopes, σ_Π^2, and in the vertical axis we have the variance of the intensities of the image, σ_I^2. In the figure we can observe the dependence of this relationship with the angular position of the source, θ_s. In figure 2 we also can observe that for small incidence angles (0-10 degrees) and small values of variance of the surface slopes, it is possible to obtain bigger values in the variance of the intensities in the image. From equation (18), we can see that this behavior is independent of any surface height power spectrum that we are analyzing, because this relation depends on the probability density function of the surface slopes and the geometry of the experiment only.

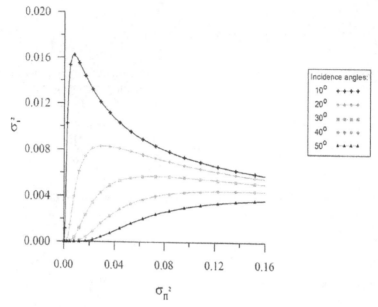

Fig. 2. Relationship between the variance of the surface slopes with the variance of the intensities in the image.

In certain cases, figure 2, if we have data corresponding to a θ_s value only, it is not possible to obtain the variance of the surface slopes, σ_Π^2, because for a value of σ_I^2 we will have two possible values of σ_Π^2. To solve this problem, it is necessary to analyze images which correspond at two or more incidence angles and to select a slope variance value which is consistent with all these data.

The relationship between σ_Π^2 and σ_ζ^2 can be derived from (Papoulis, 1981)

$$C_\Pi(\tau) = -\frac{d^2 C_\zeta(\tau)}{d\tau^2},\qquad(19)$$

if we know the correlation function of the surface heights (this will be shown in next section of this chapter). Here, $C_\zeta(\tau)$ is the correlation function of the surface heights and $C_\Pi(\tau)$ is the correlation function of the surface slopes.

2.2 Relationship between the correlation function of the intensities in the image and of the surface heights

Our analysis involves three random processes: the surface profile, $\zeta(x)$, its surface slopes, $\Pi(x)$, and the image, $I(x)$. Each process has a correlation function and it was shown (Álvarez-Borrego, 1993) that these three functions hold a relationship.

The relationship between correlation functions of the surface heights, $C_\zeta(\tau)$, and the surface slopes, $C_\Pi(\tau)$, is given by equation (19), and the relationship between $C_\Pi(\tau)$ and the correlation function of the intensities in the image, $C_I(\tau)$, is given by (Álvarez-Borrego, 1993)

$$\sigma_I^2 C_I(\tau) = \int_{-\infty}^{\infty}\int_{-\infty}^{\infty} \frac{B(\Pi_1)B(\Pi_2)}{2\pi\sigma_\Pi^2\left[1-C_\Pi^2(\tau)\right]^{1/2}} \exp\left[-\frac{\Pi_1^2 + \Pi_2^2 - 2C_\Pi(\tau)\Pi_1\Pi_2}{2\sigma_\Pi^2\left[1-C_\Pi^2(\tau)\right]}\right] d\Pi_1 d\Pi_2. \tag{20}$$

In order to achieve the inverse process, using equation (19) and equation (20), these two equations must meet certain conditions. For example, it is required that there exists one to one correspondence among the amount involved.

Using equation (19) the processed data can be numerically integrated twice, such that we obtain information of the correlation function of the surface heights, $C_\zeta(\tau)$, from the correlation function of the surface slopes, $C_\Pi(\tau)$. Although equation (20) is a more complicated expression, we cannot obtain an analytical result from it. A first integral can be analytically solved and for the second it is possible to obtain the solution by numerical integration. Resolving the first integral analytically, equation (20) can be written like

$$\sigma_I^2 C_I(\tau) = \int_a^b \frac{\sqrt{2}}{4\sigma_\Pi\sqrt{\pi}} \exp\left[-\frac{\Pi_2^2}{2\sigma_\Pi^2}\right]\left\{erf\left(\frac{b-C_\Pi(\tau)\Pi_2}{\sqrt{2\sigma_\Pi^2\left[1-C_\Pi^2(\tau)\right]}}\right) - erf\left(\frac{a-C_\Pi(\tau)\Pi_2}{\sqrt{2\sigma_\Pi^2\left[1-C_\Pi^2(\tau)\right]}}\right)\right\} d\Pi_2, \tag{21}$$

where $a = \Pi_o - \left(1+\Pi_o^2\right)\beta/4$ and $b = \Pi_o + \left(1+\Pi_o^2\right)\beta/4$.

So, a relationship between values of the correlation function of the intensities in the image, $C_I(\tau)$, and the values of the correlation function of the surface slopes takes, $C_\Pi(\tau)$, can be obtained (Figure 3). In this case, to small angles we can find higher values for the correlation function of the intensities in the image. In all the cases, the angular position of the camera or detector, θ_d, is zero and $\sigma_\Pi^2 = 0.03$. The correlation functions of figure 3 are normalized.

Also, from equation (19), it is possible to obtain the correlation function of the surface heights, $C_\zeta(\tau)$, from $C_\Pi(\tau)$ and the require inverse process to determine the correlation function of the surface heights is completed.

A theoretical variance σ_I^2 can be calculated from equation (21). We wrote in Table 1 the values of the image variance in order to normalize the correlations in figure 3 for different values for θ_s.

θ_s	σ_Π^2	σ_I^2
10	0.03	0.0119734700
20	0.03	0.0083223130
30	0.03	0.0044081650
40	0.03	0.0016988780
50	0.03	0.0004438386

Table 1. Values of the image variance in order to normalize the correlations in figure 3 for different values for θ_s.

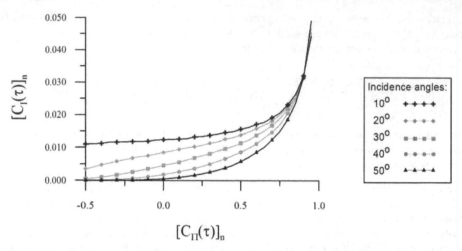

Fig. 3. Relationship between the correlation function of the surface slopes and the correlation function of the intensities in the image.

3. Geometry of the model (Gaussian case considering a variable detector angle)

A more real physical situation is shown in figure 4. The surface, $\zeta(x)$, is illuminated by a uniform incoherent source S of limited angular extent, with wavelength $\bar{\lambda}$. Its image is formed in D by an aberration-free optical system. The incidence angle θ_s is defined as the angle between the incidence angle direction and the normal to the mean surface and represents the mean angle subtended by the source S. $(\theta_d)_i$ corresponds to the angle subtended by the optical system of the detector with the normal to point i of the surface, i. e.

$$(\theta_d)_i = \tan^{-1}\left(\frac{i\Delta x}{H}\right),\tag{22}$$

where H is the height of the detector and Δx is the interval between surface points. We can see that in this more realistic physical situation, angle θ_d is changing with respect to each point in the surface. It is worth noticing that a variable θ_d does not restrict the sensor field of view.

α_i is the angle subtended between the normal to the mean surface and the normal to the slope for each i point in the surface

$$\alpha_i = \frac{\theta_s + (\theta_d)_i}{2} = \frac{\theta_s}{2} + \frac{1}{2}\tan^{-1}\left(\frac{i\Delta x}{H}\right).\tag{23}$$

The apparent diameter of the source is β. Light from the source is reflected on the surface for just one time, and, depending on the slope, the light reflected will or will not be part of the image. Thus, the image consists of bright and dark regions that we call a glitter pattern.

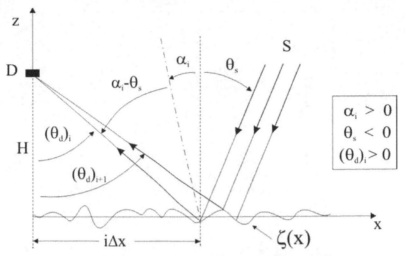

Fig. 4. Geometry of the real physical situation. Counterclockwise angles are considered as positive and clockwise angles as negative.

The glitter function can be expressed as (Álvarez-Borrego & Martín-Atienza, 2010)

$$B(\Pi_i) = rect\left[\frac{\Pi_i - \Pi_{oi}}{\left(1 + \Pi_{oi}^2\right)\frac{\beta}{2}}\right], \tag{24}$$

where

$$\Pi_{oi} - \left(1 + \Pi_{oi}^2\right)\frac{\beta}{4} \le \Pi_i \le \Pi_{oi} + \left(1 + \Pi_{oi}^2\right)\frac{\beta}{4}, \tag{25}$$

$$\Pi_i = \tan(\alpha_i), \tag{26}$$

$$\Pi_{oi} = \tan\left[\frac{\theta_s + (\theta_d)_i}{2}\right]. \tag{27}$$

The interval characterized by equation (25) defines a specular band where certain slopes generate bright spots in the image. This band has now a nonlinear slope due to the variation of $(\theta_d)_i$ with respect to each i point of the surface (Figure 5). Combining equations (25) – (27), the slope interval, where a bright spot is received by the detector, is

$$\frac{\theta_s}{2} + \frac{1}{2}\tan^{-1}\left(\frac{i\Delta x}{H}\right) - \frac{\beta}{4} \le \alpha_i \le \frac{\theta_s}{2} + \frac{1}{2}\tan^{-1}\left(\frac{i\Delta x}{H}\right) + \frac{\beta}{4}. \tag{28}$$

3.1 Relationships among the variances of the intensities in the image and surface slopes

The mean of the image μ_I may be written as (Álvarez-Borrego & Martín-Atienza, 2010)

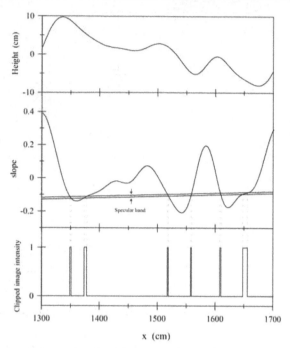

Fig. 5. All the random processes involved in our analysis. The specular band corresponds to bright regions in the image.

$$\mu_I = \langle I(x) \rangle = \int_{-\infty}^{\infty} B(\Pi_i) p(\Pi_i) d\Pi_i, \qquad (29)$$

where $B(\Pi_i)$ is the glitter function defined be equation (24). $p(\Pi_i)$ is the probability density function, where a Gaussian function is considered in one dimension. Substituting in equation (29) the expressions for $B(\Pi_i)$ and $p(\Pi_i)$, we have

$$\mu_I = \langle I(x) \rangle = \frac{1}{N} \sum_{i=1}^{N} \frac{1}{\sigma_\Pi \sqrt{2\pi}} \int_{-\infty}^{\infty} rect \left[\frac{\Pi_i - \Pi_{oi}}{(1+\Pi_{oi}^2)\frac{\beta}{2}} \right] \exp\left(-\frac{\Pi_i^2}{2\sigma_\Pi^2} \right) d\Pi_i. \qquad (30)$$

The detector angle θ_d is a function of the position x ; thus, the specular angle is a function of the distance x from the nadir point of the detector $n=0$ to the point $n=i$ (equation 22).

Defining $a_i = \Pi_{oi} - (1+\Pi_{oi}^2)\beta/4$ and $b_i = \Pi_{oi} + (1+\Pi_{oi}^2)\beta/4$, we can write

$$\mu_I = \langle I(x) \rangle = \frac{1}{N} \sum_{i=1}^{N} \frac{1}{2} \left[erf\left(\frac{b_i}{\sqrt{2}\sigma_\Pi} \right) - erf\left(\frac{a_i}{\sqrt{2}\sigma_\Pi} \right) \right]. \qquad (31)$$

The variance of the intensities in the image σ_I^2 is defined by (Álvarez-Borrego & Martín-Atienza, 2010)

$$\sigma_I^2 = \left\langle I^2(x) \right\rangle - \left\langle I(x) \right\rangle^2 = \frac{1}{N} \sum_{i=1}^{N} \int_{-\infty}^{+\infty} \left[B(\Pi_i) - \mu_I \right]^2 p(\Pi_i) d\Pi_i. \tag{32}$$

However, $B(\Pi_i) = B^2(\Pi_i)$, then $\left\langle I^2(x) \right\rangle = \left\langle I(x) \right\rangle$; therefore

$$\sigma_I^2 = \left\langle I(x) \right\rangle - \left\langle I(x) \right\rangle^2 = \mu_I (1 - \mu_I). \tag{33}$$

Substituting the equation (31) in equation (33), we have

$$\sigma_I^2 = \frac{1}{N} \sum_{i=1}^{N} \left\{ \frac{1}{2} \left[erf \left(\frac{b_i}{\sqrt{2}\sigma_\Pi} \right) - erf \left(\frac{a_i}{\sqrt{2}\sigma_\Pi} \right) \right] - \frac{1}{4N} \left[erf \left(\frac{b_i}{\sqrt{2}\sigma_\Pi} \right) - erf \left(\frac{a_i}{\sqrt{2}\sigma_\Pi} \right) \right]^2 \right\}, \tag{34}$$

which is the required relationship between the variance of the intensities in the image σ_I^2 and the variance of the surface slopes σ_Π^2.

The relationship between the variance of the surface slopes and the variances of the intensities of the image for different θ_s angles (10°-50°) is shown in figure 6 (equation 34). The detector is located as shown in figure 4 and the subtended angle by the source is $\beta = 0.68°$. When the camera detector is at H=100 m the behavior of the curves look similar to the curves shown in Álvarez-Borrego & Martín-Atienza, 2010 (figure 6a). In this case, we also can observe that, for big incidence angles (40° – 50°) and small values of variance of the surface slopes, it is possible to obtain bigger values in the variance of the intensities in the image.

If we analyze the figure 6j we can observe that σ_I^2 increases for lower θ_s values (10°-20°). These results match with the results presented by Álvarez-Borrego in 1993. Figure 6j was made considering an H=1000 m. The reason for this match is that the condition proposed by Álvarez-Borrego in 1993 considers a θ_d value constant (see figure 2). This condition is similar to have the sensor camera to an H value very high where the surface slopes values are considered almost constant.

Figure 6 shows how these relationships (σ_I^2 versus σ_Π^2) are changing while H is being bigger. Dark lines show limit extremes for θ_s of 10° and 50°. It can be seen that when H is increasing to 200 m the line of 50° starts to decay and start to cross with the others. In so far as H goes up, the lines, with larger θ_s go down until the order of the curves change. The explanation for this is very simple: if the camera stays at H=100 m, it will receive more reflection of light at large θ_s, because the geometry of reflection. When H increases, the camera will receive less light reflection of large incidence angles but will have more light reflection for small incidence angles. Therefore, when the camera is at a larger height, will have more reflection from light incidence angles smaller than light of larger incidence angles. Thus we can say that the results presented by Álvarez-Borrego in 1993, Cureton *et al.*, 2007 and Álvarez-Borrego & Martín-Atienza in 2010 are correct for the Gaussian case.

In certain cases, if we have data corresponding to one θ_s value, it is not possible to obtain a single value for the variance of the surface slopes σ_Π^2. To solve this problem, it is necessary to analyze images which correspond at two or more incidence angles and to select a slope variance value which is consistent with all these data (Álvarez-Borrego, 1995).

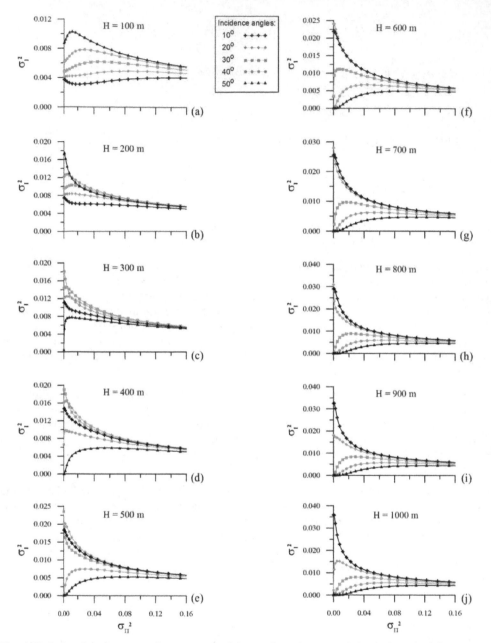

Fig. 6. Relationship between the variance of the surface slopes and the variance of the intensities of the image for different H values.

From equation (34), we can see that this relation depends on the probability density function of the surface slopes and the geometry of the experiment only.

3.2 Relationship between the correlation functions of the intensities in the image and of the surface slope

The relationship between the correlation function of the surface slopes $C_\Pi(\tau)$ and the correlation functions of the intensities in the image $C_I(\tau)$ is given by

$$\sigma_I^2 C_I(\tau) = \frac{1}{N}\sum_{i=1}^{N}\int_{-\infty}^{\infty}\frac{1}{N}\sum_{i=1}^{N}\int_{-\infty}^{\infty} B(\Pi_{1i})B(\Pi_{2i})p(\Pi_{1i},\Pi_{2i})d\Pi_{1i}\,d\Pi_{2i}, \tag{35}$$

where $p(\Pi_{1i},\Pi_{2i})$ is defined by

$$p(\Pi_{1i},\Pi_{2i}) = \frac{1}{2\pi\sigma_\Pi^2\left[1-C_\Pi^2(\tau)\right]^{1/2}}\exp\left[-\frac{\Pi_{1i}^2 - 2C_\Pi(\tau)\Pi_{1i}\Pi_{2i} + \Pi_{2i}^2}{2\sigma_\Pi^2\left(1-C_\Pi^2(\tau)\right)}\right]. \tag{36}$$

Although it is possible to obtain an analytical relationship for the first integral, for the second integral the process must be numeric. Thus, eq. (35) can be written like

$$\sigma_I^2 C_I(\tau) = \frac{1}{N}\sum_{i=1}^{N}\int_{a_i}^{b_i}\frac{1}{N}\sum_{i=1}^{N}\frac{\sqrt{2}}{4\sigma_\Pi\sqrt{\pi}}\exp\left[-\frac{\Pi_2^2}{2\sigma_\Pi^2}\right]\left\{erf\left(\frac{b_i - C_\Pi(\tau)\Pi_2}{\sqrt{2\sigma_\Pi^2\left[1-C_\Pi^2(\tau)\right]}}\right) - erf\left(\frac{a_i - C_\Pi(\tau)\Pi_2}{\sqrt{2\sigma_\Pi^2\left[1-C_\Pi^2(\tau)\right]}}\right)\right\}d\Pi_2, \tag{37}$$

where $a_i = \Pi_{oi} - \left(1+\Pi_{oi}^2\right)\beta/4$ and $b_i = \Pi_{oi} + \left(1+\Pi_{oi}^2\right)\beta/4$.

In order to avoid computer memory problems, the 16384 data point profile was divided into into a number of consecutive intervals. The value of θ_d varies point to point in the profile. For each interval and for each θ_s value, the relationship between the correlation functions $C_I(\tau)$ and $C_\Pi(\tau)$ was calculated. Then, the several computed relationships for each θ_s value were averaged.

In this case we used a value of $\sigma_\Pi^2 = 0.03$. The correlation function of the intensities in the image is not normalized. Similar to the behavior of the variances, when H increases the behavior of the curves have a similar process. A theoretical variance σ_I^2 can be calculated from equation (37). We wrote in Table 2 the values of the image variance in order to normalize the correlations in figure 7 for different values for θ_s and H (100, 500, 1000 and 5000 m).

4. Geometry of the model (Non-Gaussian case considering a variable detector angle)

The model, considering θ_d as variable, is shown in figure 4. We think this is a more realistic situation.

4.1 Relationships among the variances of the intensities in the image and surface slopes considering a non-Gaussian probability density function

The mean of the image μ_I may be written as (Álvarez-Borrego & Martín-Atienza, 2010):

H	θ_s	σ_Π^2	σ_I^2
100	10	0.03	0.00003160564
100	20	0.03	0.00005271762
100	30	0.03	0.00014855790
100	40	0.03	0.00058990210
100	50	0.03	0.00195377600
500	10	0.03	0.00015853820
500	20	0.03	0.00023902050
500	30	0.03	0.00043911520
500	40	0.03	0.00107317300
500	50	0.03	0.00269619900
1000	10	0.03	0.00031712280
1000	20	0.03	0.00047002010
1000	30	0.03	0.00078709770
1000	40	0.03	0.00161060600
1000	50	0.03	0.00344703200
5000	10	0.03	0.00158160000
5000	20	0.03	0.00228022000
5000	30	0.03	0.00332568200
5000	40	0.03	0.00498063700
5000	50	0.03	0.00723998800

Table 2. Values of the image variance in order to normalize the correlations in figure 7 for different values for θ_s and H.

$$\mu_I = \langle I(x) \rangle = \frac{1}{N} \sum_{i=1}^{N} \int_{-\infty}^{+\infty} B(\Pi_i) p(\Pi_i) d\Pi_i \qquad (38)$$

where $B(\Pi_i)$ is the glitter function defined by equation (24). $p(\Pi_i)$ is the probability density function, where a non-Gaussian function is considered in one dimension (Cureton, 2010)

$$p(\Pi_i) = \frac{1}{\sigma_\Pi \sqrt{2\pi}} \exp\left(-\frac{\Pi_i^2}{2\sigma_\Pi^2}\right) \cdot \left[1 + \frac{1}{6} \lambda_\Pi^{(3)} \left\{ \left(\frac{\Pi_i}{\sigma_\Pi}\right)^3 - 3\left(\frac{\Pi_i}{\sigma_\Pi}\right) \right\} + \frac{1}{24} \lambda_\Pi^{(4)} \left\{ \left(\frac{\Pi_i}{\sigma_\Pi}\right)^4 - 6\left(\frac{\Pi_i}{\sigma_\Pi}\right)^2 + 3 \right\} \right], \quad (39)$$

where $\lambda_\Pi^{(3)}$ is the skewness, $\lambda_\Pi^{(4)}$ is the kurtosis and σ_Π is the standard deviation of the surface slopes.

Substituting in equation (38) the expressions for $B(\Pi_i)$ and $p(\Pi_i)$, we have

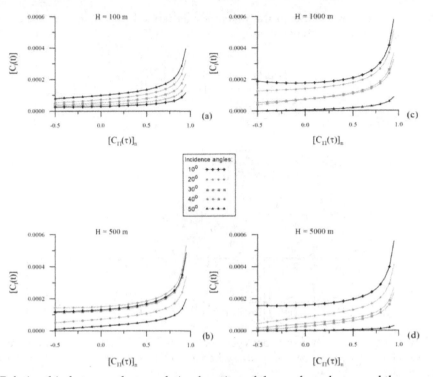

Fig. 7. Relationship between the correlation function of the surface slopes and the correlation function of the intensities in the image.

$$\mu_I = \frac{1}{N}\sum_{i=1}^{N}\frac{1}{\sigma_\Pi\sqrt{2\pi}}\int_{-\infty}^{+\infty} rect\left[\frac{\Pi_i - \Pi_{oi}}{\left(1+\Pi_{oi}^2\right)\frac{\beta}{2}}\right]\exp\left(-\frac{\Pi_i^2}{2\sigma_\Pi^2}\right)\cdot\left[1+\frac{1}{6}\lambda_\Pi^{(3)}\left\{\left(\frac{\Pi_i}{\sigma_\Pi}\right)^3 - 3\left(\frac{\Pi_i}{\sigma_\Pi}\right)\right\}\atop +\frac{1}{24}\lambda_\Pi^{(4)}\left\{\left(\frac{\Pi_i}{\sigma_\Pi}\right)^4 - 6\left(\frac{\Pi_i}{\sigma_\Pi}\right)^2 + 3\right\}\right]d\Pi_i. \quad (40)$$

The detector angle θ_d is a function of the position x, thus, the specular angle is a function of the distance x from the nadir point of the detector, $n = 0$, to the point $n = i$ (see equation (22)).

Writing again $a_i = \Pi_{oi} - \left(1+\Pi_{oi}^2\right)\beta/4$ and $b_i = \Pi_{oi} + \left(1+\Pi_{oi}^2\right)\beta/4$, we can write

$$\mu_I = \frac{1}{N}\sum_{i=1}^{N}\left\{\begin{array}{l}\left[erf\left(\frac{b_i}{\sqrt{2}\sigma_\Pi}\right) - erf\left(\frac{a_i}{\sqrt{2}\sigma_\Pi}\right)\right]\cdot\left[\frac{1}{2}+\frac{1}{8}\lambda_\Pi^{(4)}\left(1-3\sigma_\Pi^2\right)\right]+\\ +\exp\left(-\frac{a_i^2}{2\sigma_\Pi^2}\right)\cdot\left[\frac{\lambda_\Pi^{(3)}}{6\sqrt{2\pi}\sigma_\Pi^2}\left(a_i^2 - \sigma_\Pi^2\right)+\frac{\lambda_\Pi^{(4)}a_i}{24\sqrt{2\pi}\sigma_\Pi^3}\left(a_i^2 - 3\sigma_\Pi^2\right)\right]+\\ +\exp\left(-\frac{b_i^2}{2\sigma_\Pi^2}\right)\cdot\left[\frac{\lambda_\Pi^{(3)}}{6\sqrt{2\pi}\sigma_\Pi^2}\left(\sigma_\Pi^2 - b_i^2\right)+\frac{\lambda_\Pi^{(4)}b_i}{24\sqrt{2\pi}\sigma_\Pi^3}\left(3\sigma_\Pi^2 - b_i^2\right)\right]\end{array}\right\}. \quad (41)$$

The variance of the intensities in the image σ_I^2 is defined by equation (33). Substituting equation (41) in equation (33) we have

$$
\sigma_I^2 = \frac{1}{N}\sum_{i=1}^{N}\left\{
\begin{array}{l}
\left[erf\left(\frac{b_i}{\sqrt{2}\sigma_\Pi}\right)-erf\left(\frac{a_i}{\sqrt{2}\sigma_\Pi}\right)\right]\cdot\left[\frac{1}{2}+\frac{1}{8}\lambda_\Pi^{(4)}\left(1-3\sigma_\Pi^2\right)\right]+ \\[6pt]
+\exp\left(-\frac{a_i^2}{2\sigma_\Pi^2}\right)\cdot\left[\frac{\lambda_\Pi^{(3)}}{6\sqrt{2\pi}\sigma_\Pi^2}\left(a_i^2-\sigma_\Pi^2\right)+\frac{\lambda_\Pi^{(4)}a_i}{24\sqrt{2\pi}\sigma_\Pi^3}\left(a_i^2-3\sigma_\Pi^2\right)\right]+ \\[6pt]
+\exp\left(-\frac{b_i^2}{2\sigma_\Pi^2}\right)\cdot\left[\frac{\lambda_\Pi^{(3)}}{6\sqrt{2\pi}\sigma_\Pi^2}\left(\sigma_\Pi^2-b_i^2\right)+\frac{\lambda_\Pi^{(4)}b_i}{24\sqrt{2\pi}\sigma_\Pi^3}\left(3\sigma_\Pi^2-b_i^2\right)\right]
\end{array}
\right\}-
$$

$$
-\frac{1}{N^2}\left[\sum_{i=1}^{N}
\begin{array}{l}
\left[erf\left(\frac{b_i}{\sqrt{2}\sigma_\Pi}\right)-erf\left(\frac{a_i}{\sqrt{2}\sigma_\Pi}\right)\right]\cdot\left[\frac{1}{2}+\frac{1}{8}\lambda_\Pi^{(4)}\left(1-3\sigma_\Pi^2\right)\right]+ \\[6pt]
+\exp\left(-\frac{a_i^2}{2\sigma_\Pi^2}\right)\cdot\left[\frac{\lambda_\Pi^{(3)}}{6\sqrt{2\pi}\sigma_\Pi^2}\left(a_i^2-\sigma_\Pi^2\right)+\frac{\lambda_\Pi^{(4)}a_i}{24\sqrt{2\pi}\sigma_\Pi^3}\left(a_i^2-3\sigma_\Pi^2\right)\right]+ \\[6pt]
+\exp\left(-\frac{b_i^2}{2\sigma_\Pi^2}\right)\cdot\left[\frac{\lambda_\Pi^{(3)}}{6\sqrt{2\pi}\sigma_\Pi^2}\left(\sigma_\Pi^2-b_i^2\right)+\frac{\lambda_\Pi^{(4)}b_i}{24\sqrt{2\pi}\sigma_\Pi^3}\left(3\sigma_\Pi^2-b_i^2\right)\right]
\end{array}
\right]^2
\tag{42}
$$

which is the required relationship between the variance of the intensities in the image σ_I^2 and the variance of the surface slopes σ_Π^2 when a non-Gaussian probability density function is considered.

The relationship between the variance of the surface slopes and the variances of the intensities of the image for different θ_s angles (10°-50°) is shown in figures 8 and 9 (equation 42). Figures 8 and 9 show this relationship considering the skewness and the skewness and kurtosis in the non-Gaussian probability density function respectively. We can see that the behavior of the curves looks very similar to the Gaussian case (figure 6). The values for skewness and kurtosis were taken from a Table showed by Plant (2003) from data given by Cox & Munk (1956), for a wind speed of 13.3 m/s with the wind sensor at 12.5 m on the sea surface level.

The curves including the skewness and skewness and kurtosis are little higher for small values of σ_Π^2 compared with the Gaussian case (figure 6) except when θ_s is below 40° where the Gaussian and non-Gaussian cases (considering skewness only) are inverted to small surface slope variances, and these results show that σ_I^2 increases for higher θ_s values (figures 8a and 9a). Cox & Munk (1956) reported σ_Π^2 values of 0.04 and 0.05 like maximum values of the surface slopes in the wind direction and values of 0.03 in the cross wind direction for wind speed bigger than 10 m/s. Thus, we think that in the range for σ_Π^2 from 0-0.05 the behavior of the curves look very clear and separate each one of the other (figures 8a and 9a). If we analyze the figures 8j and 9j we can observe that σ_I^2 increases for lower θ_s values (10°-20°).

Figures 8 and 9 show how these relationships (σ_I^2 versus σ_Π^2) are changing while H is being bigger, where the skewness and skewness and kurtosis are being considered. These curves have the same behavior like in the Gaussian case and the explanation for this inversion is the same as explained before.

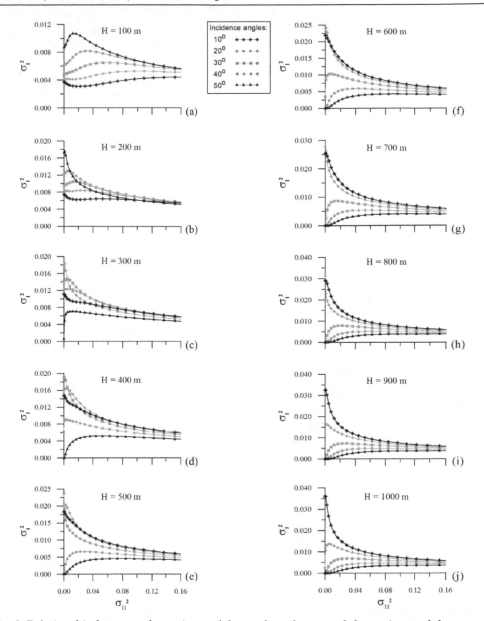

Fig. 8. Relationship between the variance of the surface slopes and the variance of the intensities of the image, for different H values considering a non-Gaussian probability density function where the skewness has been taken account only.

About the non-Gaussian case we can conclude that the main difference with the Gaussian case is the less higher values of the variance of the intensities of the image for small values of surface slope variance when θ_s is in the 40° – 50° range when H=100 m. In addition, when

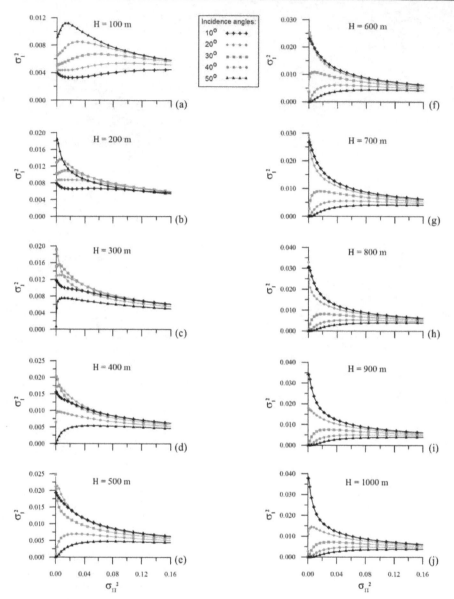

Fig. 9. Relationship between the variance of the surface slopes and the variance of the intensities of the image, for different H values considering a non-Gaussian probability density function where the skewness and kurtosis have been taken account.

H=1000 m this condition is inverted, we can find less smaller values of the variance of the intensities of the image for small values of surface slope variance when θ_s is in the 10° – 20° range. In the other angles, in both cases, it is not possible to see significant differences between the values 10° – 30° when H=100 m and 30° – 50° when H=1000 m.

4.2 Relationship between the correlation functions of the intensities in the image and of the surface slope considering a non-Gaussian probability density function

As mentioned before, our analysis involves three random processes: the surface profile $\zeta(x)$, its surface slopes $\Pi(x)$ and the image $I(x)$. Each process has a correlation function and it was shown in (Álvarez-Borrego, 1993) that these three functions are related.

The relationship between the correlation function of the surface slopes $C_\Pi(\tau)$ and the correlation function of the intensities in the image $C_I(\tau)$ is given by

$$\sigma_I^2 C_I(\tau) = \frac{1}{N}\sum_{i=1}^{N}\int_{-\infty}^{\infty}\frac{1}{N}\sum_{i=1}^{N}\int_{-\infty}^{\infty} B(\Pi_{1i})B(\Pi_{2i})p(\Pi_{1i},\Pi_{2i})d\Pi_{1i}\,d\Pi_{2i}, \tag{43}$$

where $p(\Pi_{1i},\Pi_{2i})$ is defined by (Cureton, 2010)

$$p(\Pi_{1i},\Pi_{2i}) = \frac{1}{2\pi\sigma_\Pi^2\left[1-C_\Pi^2(\tau)\right]^{1/2}}\exp\left[-\frac{\Pi_{1i}^2 - 2C_\Pi(\tau)\Pi_{1i}\Pi_{2i} + \Pi_{2i}^2}{2\sigma_\Pi^2\left(1-C_\Pi^2(\tau)\right)}\right]\times$$

$$\left\{1+\frac{1}{6}\left\{\begin{array}{l}\lambda_\Pi^{(30)}\left[\left(\frac{\Pi_{1i}}{\sigma_\Pi}\right)^3 - 3\sigma_\Pi^2\left(\frac{\Pi_{1i}}{\sigma_\Pi}\right)\right]+ \\[2mm] 3\lambda_\Pi^{(21)}\left[\left(\frac{\Pi_{1i}}{\sigma_\Pi}\right)^2\left(\frac{\Pi_{2i}}{\sigma_\Pi}\right) - \sigma_\Pi^2\left(\frac{\Pi_{2i}}{\sigma_\Pi}\right) + 2\sigma_\Pi^2 C_\Pi(\tau)\left(\frac{\Pi_{1i}}{\sigma_\Pi}\right)\right]+ \\[2mm] 3\lambda_\Pi^{(12)}\left[\left(\frac{\Pi_{1i}}{\sigma_\Pi}\right)\left(\frac{\Pi_{2i}}{\sigma_\Pi}\right)^2 - \sigma_\Pi^2\left(\frac{\Pi_{1i}}{\sigma_\Pi}\right) + 2\sigma_\Pi^2 C_\Pi(\tau)\left(\frac{\Pi_{2i}}{\sigma_\Pi}\right)\right]+ \\[2mm] \lambda_\Pi^{(03)}\left[\left(\frac{\Pi_{2i}}{\sigma_\Pi}\right)^3 - 3\sigma_\Pi^2\left(\frac{\Pi_{2i}}{\sigma_\Pi}\right)\right]\end{array}\right\}\right\}, \tag{44}$$

where $\lambda_\Pi^{(03)}$ and $\lambda_\Pi^{(30)}$ are the skewness, $\lambda_\Pi^{(12)}$ and $\lambda_\Pi^{(21)}$ are the relationship between the moments of Π_{1i} and Π_{2i}.

Although it is possible to obtain an analytical relationship for the first integral, for the second integral the process must be numeric. Thus, equation (43) can be written like

$$\sigma_I^2 C_I(\tau) = \frac{1}{N}\sum_{i=1}^{N}\int_{a_i}^{b_i}\frac{1}{N}\sum_{i=1}^{N}\exp\left(-\frac{\Pi_{2i}}{2\sigma_\Pi^2}\right)\times\left\{\begin{array}{l}\exp\left[-(ub_i+v\Pi_{2i})^2\right]\times\left(A_1\Pi_{2i}^2 + B_1b_i\Pi_{2i} + C_1\right) \\[1mm] +\exp\left[-(ua_i+v\Pi_{2i})^2\right]\times\left(A_2\Pi_{2i}^2 + B_2a_i\Pi_{2i} + C_2\right) \\[1mm] +\left[erf(ub_i+v\Pi_{2i}) - erf(ua_i+v\Pi_{2i})\right]\times\left(A_3\Pi_{2i}^3 + B_3\Pi_{2i} + C_3\right)\end{array}\right\}d\Pi_{2i}, \tag{45}$$

where

$$u = \frac{1}{\sqrt{2\sigma_\Pi^2\left[1-C_\Pi^2(\tau)\right]}},$$

$$v = \frac{-C_\Pi(\tau)}{\sqrt{2\sigma_\Pi^2\left[1-C_\Pi^2(\tau)\right]}},$$

$$A_1 = -\frac{\sqrt{1-C_\Pi^2(\tau)}}{12\pi\sigma_\Pi^3}\left(C_\Pi^2(\tau)\lambda_\Pi^{(30)} + 3C_\Pi(\tau)\lambda_\Pi^{(21)} + 3\lambda_\Pi^{(12)}\right) = -A_2,$$

$$B_1 = -\frac{\sqrt{1-C_\Pi^2(\tau)}}{12\pi\sigma_\Pi^3}\left(C_\Pi(\tau)\lambda_\Pi^{(30)} + 3\lambda_\Pi^{(21)}\right) = -B_2,$$

$$C_1 = -\frac{\sqrt{1-C_\Pi^2(\tau)}}{12\pi\sigma_\Pi^3}\left[\left(b_i^2 + 2\sigma_\Pi^2\left[1-C_\Pi^2(\tau)\right] - 3\sigma_\Pi^4\right)\lambda_\Pi^{(30)} + 6\sigma_\Pi^4 C_\Pi(\tau)\lambda_\Pi^{(21)} - 3\sigma_\Pi^4\lambda_\Pi^{(12)}\right],$$

$$C_2 = \frac{\sqrt{1-C_\Pi^2(\tau)}}{12\pi\sigma_\Pi^3}\left[\left(a_i^2 + 2\sigma_\Pi^2\left[1-C_\Pi^2(\tau)\right] - 3\sigma_\Pi^4\right)\lambda_\Pi^{(30)} + 6\sigma_\Pi^4 C_\Pi(\tau)\lambda_\Pi^{(21)} - 3\sigma_\Pi^4\lambda_\Pi^{(12)}\right],$$

$$A_3 = \frac{\sqrt{2}}{24\sqrt{\pi}\sigma_\Pi^4}\left(C_\Pi^3(\tau)\lambda_\Pi^{(30)} + 3C_\Pi^2(\tau)\lambda_\Pi^{(21)} + 3C_\Pi(\tau)\lambda_\Pi^{(12)} + \lambda_\Pi^{(03)}\right),$$

$$B_3 = \frac{\sqrt{2}}{8\sqrt{\pi}}\left[C_\Pi(\tau)\left(\frac{1-C_\Pi^2(\tau)}{\sigma_\Pi^2}-1\right)\lambda_\Pi^{(30)} + \left(2C_\Pi^2(\tau)+\frac{1-C_\Pi^2(\tau)}{\sigma_\Pi^2}-1\right)\lambda_\Pi^{(21)} + C_\Pi(\tau)\lambda_\Pi^{(12)} - \lambda_\Pi^{(03)}\right],$$

$$C_3 = \frac{\sqrt{2}}{4\sqrt{\pi}\sigma_\Pi}.$$

Figure 10 shows graphically the relationship between the normalized correlation function of the surface slopes $\left[C_\Pi(\tau)\right]_n$ and the normalized correlation function of the intensities of the image $\left[C_I(\tau)\right]_n$. In this case a $\sigma_\Pi^2 = 0.03$ was used. When H increases the behavior of the curves have a similar process like the variance curves.

When H=100 m (Figure 10a) the behavior of the curves for θ_s of 10° – 20° have an "unusual" behavior for low surface slope variances when compared with Gaussian case. This is because the inversion of the curves starts to lower values of H. In order to avoid memory computer problems, the 16384 data points profile was divided into a number of consecutive intervals. The value of θ_d varies point to point in the profile. For each interval and for each θ_s value, the relationship between the correlation functions $C_I(\tau)$ and $C_\Pi(\tau)$ was calculated. Then, all the computed relationships for each θ_s value were averaged.

A theoretical variance σ_I^2 can be calculated from equation (45). We wrote in Table 3 the values of the image variance in order to normalize the correlations in figure 10 for different values for θ_s and H (100, 500, 1000 and 5000 m).

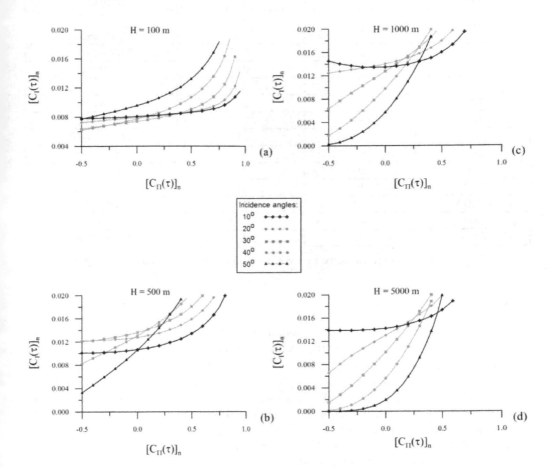

Fig. 10. Relationship between the correlation function of the surface slopes and the correlation function of the intensities in the image. The curves correspond to different values of θ_s.

H	θ_s	σ_{Π}^2	σ_I^2
100	10	0.03	0.003126364
100	20	0.03	0.004354971
100	30	0.03	0.006071378
100	40	0.03	0.008187813
100	50	0.03	0.009875824
500	10	0.03	0.012038690
500	20	0.03	0.011886750
500	30	0.03	0.009668245
500	40	0.03	0.006645083
500	50	0.03	0.003959459
1000	10	0.03	0.012945720
1000	20	0.03	0.010339930
1000	30	0.03	0.006902623
1000	40	0.03	0.004036960
1000	50	0.03	0.002067475
5000	10	0.03	0.011358240
5000	20	0.03	0.007713670
5000	30	0.03	0.004572885
5000	40	0.03	0.002406005
5000	50	0.03	0.001022463

Table 3. Values of the image variance in order to normalize the correlations in figure 10 for different values for θ_s and H.

5. Conclusions

We derive the variance of the surface heights from the variance of the intensities in the image via remote sensing considering a glitter function given by equation (12) when the geometry consider a detector angle of $\theta_d = 0^o$, and considering a glitter function given by the equation (24) considering a geometrically improved model with variable detector line of sight angle, given by figure 4. In this last case, we consider Gaussian statistics and non-Gaussian statistics. We derive the variance of the surface slopes from the variance of the intensities of remote sensed images for different H values. In addition, we discussed the determination of the correlation function of the surface slopes from the correlation function of the image intensities considering Gaussian and non-Gaussian statistics.

Analyzing the variances curves for Gaussian and non-Gaussian case it is possible to see the behavior of the curves for different incident angles when H increases. This behavior agrees with the results presented by Álvarez-Borrego (1993) and Geoff Cureton et al. 2007, and Álvarez-Borrego and Martin-Atienza (2010) for the Gaussian case.

These new results solve the inverse problem when it is necessary to analyze the statistical of a real sea surface via remote sensing using the image of the glitter pattern of the marine surface.

6. Acknowledgments

This work was partially supported by CONACyT with grant No. 102007 and SEP-PROMET/103.5/10/5021 (UABC-PTC-225).

7. References

Álvarez-Borrego, J. (1987). Optical analysis of two simulated images of the sea surface. *Proceedings SPIE International Society of the Optical Engineering*, Vol.804, pp.192-200, ISSN 0277-786X

Álvarez-Borrego, J. (1993). Wave height spectrum from sun glint patterns: an inverse problem. *Journal of Geophysical Research*, Vol.98, No.C6, pp. 10245-10258, ISSN 0148-0227

Álvarez-Borrego, J. (1995). Some statistical properties of surface heights via remote sensing. *Journal of Modern Optics*, Vol.42, No.2, pp. 279-288, ISSN 0950-0340

Álvarez-Borrego, J. & Machado M. A. (1985). Optical analysis of a simulated image of the sea surface. *Applied Optics*, Vol.24, No.7, pp. 1064-1072, ISSN 1559-128X

Álvarez-Borrego, J. & Martin-Atienza, B. (2010). An improved model to obtain some statistical properties of surface slopes via remote sensing using variable reflection angle. *IEEE Transactions on Geoscience and Remote Sensing*, Vol.48, No.10, pp. 3647-3651, ISSN 0196-2892

Bréon, F. M. & Henrist N. (2006). Spaceborn observations of ocean glint reflectance and modeling of wave slope distributions. *Journal Geophysical Research*, Vol.111, CO6005, ISSN 0148-0227

Chapman, R. D. & Irani G. B. (1981). Errors in estimating slope spectra from wave images. *Applied Optics*, Vol.20, No.20, pp. 3645-3652, ISSN 1559-128X

Chapron, B.; Vandemark D. & Elfouhaily T. (2002). On the skewness of the sea slope probability distribution. *Gas Transfer at Water Surfaces*, Vol.127, pp. 59-63, ISSN 0875909868

Cox, C. & Munk W. (1954a). Statistics of the sea surface derived from sun glitter. *Journal Marine Research*, Vol.13, No.2, pp. 198-227, ISSN 0022-2402

Cox, C. & Munk W. (1954b). Measurements of the roughness of the sea surface from photographs of the Sun's glitter. *Journal of the Optical Society of America*, Vol.24, No.11, pp. 838-850, ISSN 1084-7529

Cox, C. & Munk W. (1955). Some problems in optical oceanography. *Journal of Marine Research*, Vo.14, pp. 63-78, ISSN 0022-2402

Cox, C. & Munk. W. (1956). Slopes of the sea surface deduced from photographs of sun glitter. *Bulletin of the Scripps Institution of Oceanography*, Vol.6, No.9, pp. 401-488

Cureton, G. P. (2010). *Retrieval of nonlinear spectral information from ocean sunglint*. PhD thesis, Curtin University of Technology, Australia, March

Cureton, G. P.; Anderson, S. J.; Lynch, M. J. & McGann, B. T. (2007). Retrieval of wind wave elevation spectra from sunglint data. *IEEE Transactions on Geoscience and Remote Sensing*, Vol.45, No.9, pp. 2829-2836, ISSN 0196-2892

Fuks, I. M. & Charnotskii, M. I. (2006). Statistics of specular points at a randomly rough surface. *Journal of the Optical Society of America, Optical Image Science*, Vol.23, No.1, pp. 73-80, ISSN 1084-7529

Gaskill, J. D. (1978). *Linear systems, Fourier transform, and optics.* John Wiley & Sons. ISBN 0-471-29288-5, New York, USA

Longuet-Higgins, M. S. (1962). The statistical geometry of random surfaces. *Proceedings Symposium Applied Mathematics 1960 13th Hydrodynamic Instability,* pp. 105-143

Longuet-Higgins, M. S.; Cartwright, D. E. & Smith, N. D. (1963). Observations of the directional spectrum of sea waves using the motions of a floating buoy, In: *Ocean Wave Spectra,* Prentice-Hall, Englewood Cliffs, N. J. (Ed.), 111-136

Munk, W. (2009). An inconvenient sea truth: spread, steepness, and skewness of surface slopes. *Annual Review of Marine Sciences,* Vol.1, pp. 377-415, ISSN 1941-1405

Papoulis, A. (1981). *Probability, Random Variables, and Stochastic Processes,* chapter 9, McGraw-Hill, ISBN 0-07-119981-0, New York, USA

Peppers, N. & Ostrem, J. S. (1978). Determination of wave slopes from photographs of the ocean surface: A new approach. *Applied Optics,* Vol.17, No.21, pp. 3450-3458, ISSN 1559-128X

Plant, W. J. (2003). A new interpretation of sea-surface slope probability density functions. *Journal of Geophysical Research,* Vol.108, No.C9, 3295, ISSN 0148-0227

Stilwell, D. Jr. (1969). Directional energy spectra of the sea from photographs. *Journal of Geophysical Research,* Vol.74, No.8, pp. 1974-1986, ISSN 0148-0227

Stilwell, D. Jr. & Pilon, R. O. (1974). Directional spectra of surface waves from photographs. *Journal of Geophysical Research,* Vol.79, No.9, pp.1277-1284, ISSN 0148-0227

2

Characterizing Forest Structure by Means of Remote Sensing: A Review

Hooman Latifi

Dept. of Remote Sensing and Landscape Information Systems, University of Freiburg
Germany

1. Introduction

1.1 Forest structural attributes

Forest management comprises of a wide range of planning stages and activities which are highly variable according to the goals and strategies being pursued. Furthermore, those activities often include a requirement for description of condition and dynamics of forests (Koch et al., 2009). Forest ecosystems are often required to be described by a set of general characteristics including composition, function, and structure (Franklin, 1986). Composition is described by presence or dominance of woody species or by relative indices of biodiversity. Forest functional characteristics are related to issues like types and rates of processes such as carbon sequestration. Apart from them, the physical characteristics of forests are essential to be expressed. This description is often accomplished under the general concept of forest structure. However, the entire above-mentioned characteristics are required for timber management/procurement practices, as well as for mapping forests into smaller units or compartments.

The definition by (Oliver & Larson, 1996) can be referred to as one of the basic ones, in which forest structure is defined as 'the physical and temporal distribution of trees in a forest stand'. This definition encompasses a set of indicators including species distribution, vertical and horizontal spatial patterns, tree size, tree age and/or combinations of them. Yet, a more geometrical representation of forest stand was previously presented by e.g. (Franklin, 1986) or later by (Kimmins, 1996). They defined stand structure as the vertical and horizontal association of stand elements. Despite the differences between the above-mentioned definitions, they were later used as basis to derive further representative structural indicators which are mainly derived based on the metrics such as diameter at breast height (DBH). The reason is the straightforwardness and (approximately) unbiasedness of its measurement in terrestrial surveys (Stone & Porter, 1998). The interest in applying geometric derivations e.g. standing volume and aboveground biomass was later accomplished thanks to the progresses in computational facilities and simulation techniques. Those attributes are still of great importance to describe forest stand structure. Nevertheless, (McElhinny et al., 2005) stated that the structural, functional and compositional attributes of a stand are highly interdependent and thus cannot be easily divided to such main categories, since the attributes from either of the groups can be considered as alternatives to each other. Thus a new category was created, according to which the structural attributes were in a group comprising of measures such as abundance (e.g. dead wood volume), size variation (e.g. variation in DBH) and spatial variation (e.g. variation of distance to a nearest neighbour (Table 1) (McElhinny et al., 2005).

Though canopy cover i.e. the vertical projection of tree crowns is often referred to as an attribute characterizing the distribution of forest biomass, there are further attributes such as basal area, standing timber volume and the height of overstory which are considered as the more representative descriptors of forest biomass. Moreover, a combination of those attributes (especially in accordance with species composition) is also reported by e.g. (Davey, 1984) to represent the biomass and vertical complexity of the stands.

Forest stand element	Structural attribute
Foliage	Foliage height diversity
	Number of strata
	Foliage density within different strata
Canopy cover	Canopy cover
	Gap size classes
	Average gap size and the proportion of canopy in gaps
	Proportion of crowns with dead and broken tops
Tree diameter	Diameter at Breast Height (DBH)
	standard deviation of DBH
	Diameter distribution
	Number of large trees
Tree height	Height of overstorey
	Standard deviation of tree height
	Height classes richness
Tree spacing	Clark - Evans and Cox indices, percentage of trees in clusters
	Stem count per ha
Stand biomass	Stand basal area
	Standing volume
	Biomass
Tree species	Species diversity and/or richness
	Relative abundance of key species
Overstorey vegetation	Shrub height
	Shrub cover
	Total understorey cover
	Understorey richness
	Saplings (shade tolerant) per ha
Dead wood	Number, volume or basal area of stags
	Volume of coarse woody debris
	Log volume by decay or diameter classes
	Coefficient of variation of log density

Table 1. Broadly-investigated forest structural attributes, grouped under the stand element under description (after (McElhinny et al., 2005).

In addition, stem count has also been reported as an important indicator of e.g. felled logs or trees with hollows, since they offer potential habitats for the wildlife ((Acker et al., 1998), (McElhinny et al., 2005)). Thus, the frequency of larger stems is considered of more significance as a descriptor of stand structure, as it can mainly characterize the older and mature stems within the overstory of the stands. This attribute (stem count of older trees) has been already studied by e.g. (Van Den Meersschaut & Vandekerkhove, 1998) as a structural feature to distinguish the old-growth stands from the early stages of succession. Although some studies combined stem count with measures of diameter distribution e.g. (Tyrrell & Crow, 1994), some studies e.g. (Uuttera et al., 1997) did not suggest diameter distribution to be essentially helpful for describing forest structure, as comparing the diameter distributions from different stands bears some degree of sophistication.

All in all, the structural features of forest stands, as stated above, are entirely considered to be useful when describing the horizontal and vertical complexity of the forested areas. However, a relatively limited number of those attributes have been attempted to be modelled by means of remote sensing. Only a few studies have focused on other spatially-meaningful characteristics such as gaps or coarse woody debris e.g. (Pesonen et al., 2008) which have been almost entirely conducted across Scandinavian boreal forests, where the homogenous composition, single-story stands (consisting mainly of coniferous species) and topographically-gentle landscape minimise the problems of characterizing more complex descriptors of forest structure.

Since earth observation data has been applied for forestry applications, the majority of modelling tasks have been accomplished by focusing on standing timber volume, stand height, aboveground biomass (AGB), stem count, and diameter distribution as structural attributes. Whereas some compositional characteristics such as species richness/abundance have also been considered as forest structural attributes (Table 1), this article will not review their related literature, as they follow, in the scope of remote sensing, entirely different methodological strategies and thus require separate review studies with more concentration on pixel-based analysis and spectrometry.

Estimation of AGB in forest is obviously of a great importance. The rationale is straightforward: As the available stocks of fossil fuels gradually diminish and the environmental effects of climate change increasingly emerge, a wide range of stakeholders including political, economical and industrial sectors endeavour to adjust to the consequences and adapt the existing energy supply to the ongoing developments. To this aim, a vital step is the assessment of the potential renewable energy sources such as biomass. Germany can be referred as an example, in which approximately 17 million ha of farmland and 11 million ha of forest are potentially reported to be available as bioenergy sources (BMU, 2009). Moreover, according to the results of the German National Forest Inventory, around 1.0 to 1.5 percent of the country's primary energy demand (20 and 25 million m^3) in 2006 was supplied by timber products. The current models even confirm that an additional 12 to 19 million $m^3\ year^{-1}$ of timber can be sustainably used for energy production. This can in turn justify the necessity of an efficient monitoring system for assessing the potential biomass resources in regional and local levels.

1.2 Remote sensing for retrieval of forest attributes

In Recent years the general interest in forests and the environmental-related issues has exceedingly increased. This, together with the ongoing technological developments such as improved data acquisition and computing techniques, has fostered progresses in forest monitoring processes, where the assessment of environmental processes has been enabled to be carried out by means of advanced methods such as intensive modelling and simulations

(Guo, 2005). As described above, assessment and mapping of forest attributes have followed a similar progress as an essential prerequisite for forest management practices.

Information within each forest management unit (e.g. sample plots or segments characterising forest stands) often includes attributes that are measured using direct measurement (e.g. field-based surveys) and indirect measurement (e.g. mathematical derivations and modelled/simulated data). Detailed ground-based survey of each unit is reported by e.g. (LeMay & Temesgen, 2005) to be unlikely, particularly in large-area surveys dealing with limited financial resources or in the inventory of small areas, when those areas are under private ownerships. Such areas are usually associated with financial problems for regular plot-based surveys. However, the plot-based inventory data are considered as being essential as representatives of the current forest inventory or as model inputs to project the future conditions. In order to overcome the mentioned limitations in regular terrestrial surveys, one approach is to combine field measurements with airborne and spaceborne remotely-sensed data to retrieve the required information. This can in turn offer combined practical applications of the field data that represent the detailed information on the ground supported by those data which represent the spatial, spectral and temporal merits of satellite or airborne sensors (Figure 1).

Based on this potential cost-effective implications, a range of applications have been developed which enable one to pursue different natural resource planning objectives including retrieval of forest structural attributes. Amongst the most important international forest mapping projects using earth observation data, GMES (Global Monitoring for Environment and Security), TREES (Tropical Ecosystem Environment Observation by Satellite) and FRA (Forest Resource Assessment) can be highlighted (Koch, 2010). Depending on the specific application, the required level of details and especially the required accuracy of output information, variety of remotely sensed data sources can be potentially applied including a wide range of optical data (broadband multispectral and narrowband hyperspectral imagery), Radio Detection and Ranging (RADAR) and recently Light Detection and Ranging (LiDAR) data. Each one of those data sources has been proved to bear potentials and advantages for forestry applications. Whereas LiDAR instruments facilitate collecting detailed information which accurately captures the three-dimensional structure of the earth surface, RADAR data enable one to overcome common atmospheric and shadow effects which often occur in forested areas. Broadband optical data is able to reflect the general spectral responses of natural and manmade objects including vegetation cover over a big scene, while imaging spectroscopy data has been shown to provide a rich source of spectral information for various applications e.g. tree species classification.

Compared to other sources of data, LiDAR data has been successfully validated for studying the structure of forested areas. Laser altimetry is an active remote sensing technology that determines ranges by taking the product of the speed of light and the time required for an emitted laser to travel to a target object. The elapsed time from when a laser is emitted from a sensor and intercepts an object can be measured using either pulsed ranging (where the travel time of a laser pulse from a sensor to a target object is recorded) or continuous wave ranging (where the phase change in a transmitted sinusoidal signal produced by a continuously emitting laser is converted into travel time) (Wehr & Lohr, 1999). LiDAR is capable of providing both horizontal and vertical information with the horizontal and vertical sampling. The quality of sampling depends on the type of LiDAR system used and on whether it is discrete return or full waveform LiDAR system (Lim et al., 2003).

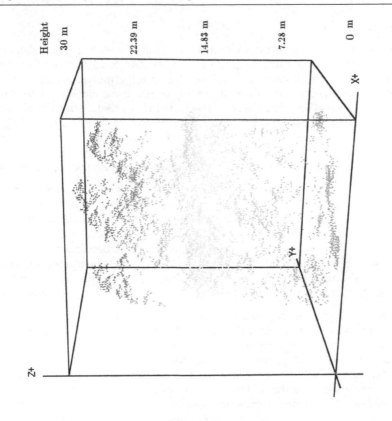

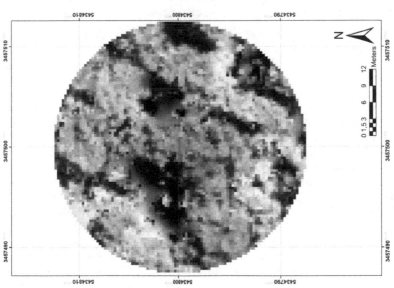

Fig. 1. An example of false colour composite from Colour Infrared (CIR) aerial images (Left) and normalized first-pulse LiDAR point cloud (right) demonstrating a circular forest inventory plot($452.4\ m^2$) in a test site in Karlsruhe, Germany.

1.3 Modelling issues

When the aim is to assess the forest attributes by means of remote sensing data, one may note, again, the importance of estimating forest biomass. (Koch, 2010) states that three main factors of forest height, forest closure and forest type are the most meaningful descriptors for AGB. Remote sensing-derived information from the above-mentioned sources will enable one to successfully assess those three factors which can in turn result in reasonable estimation of forest AGB. By using those auxiliary data as descriptors of forest structure (e.g. AGB), Statistical methods are used to model the forest stand attributes in different scales including regional, stand and individual tree levels. So far, the modelling process has been mostly accomplished by means of parametric regression modelling of the response attributes.

Parametric models generally come with strong assumptions of distributions for the parameters and variables which sometimes may not be met by the data. The application of those models is normally subjected to the scientific, technological, and logistic conditions which constrain their application in many cases (Cabaravdic, 2007). A parametric fitting can yield highly biased models resulted from the possible misspecification of the unknown density function (e.g. (Härdle, 1990)). Nevertheless, those modelling procedures have been widely used for building models of forest stand and single tree attributes by several studies (e.g.(Næsset, 2002), (Breidenbach et al., 2008), (Korhonen et al., 2008), and (Straub et al., 2009)).

In contrast, the so called âĂIJnonparametric methodsâĂİ allow for more flexibility in using the unknown regression relationships. (Härdle, 1990) and (Härdle et al., 2004) discussed four main motivations to start with nonparametric models: 1)they provide flexibility to explore the relationships between the predictor and response variables, 2)they enable predictions which are independent from reference to a fixed parametric model, 3)they can help to find false observations by studying the influence of isolated points, and 4) they can be considered as versatile methods for imputing missing values or interpolations between neighbouring predictor values. However, they require larger sample sizes than parametric counterparts, as the underlying data in a nonparametric approach simultaneously serves as the model input.

The nonparametric methods include a wide range of model-fitting approaches such as smoothing methods (e.g. kernel smoothing, k-nearest neighbour, splines and orthogonal series estimators), Generalized Additive Models (GAMs) and models based on classification and regression trees (CARTs). The k-nearest neighbour (k-NN) method is known as a group of mostly-applied nonparametric methods. In k-NN method, the value of the response variable(s) of interest on a specific target unit is modelled as a weighted average of the values of the most similar observation(s) in its neighbourhood. The neighbour(s) are defined within an n-dimensional feature space consisted of potentially-relevant predictor variables. The chosen neighbour(s) are selected based on a criterion which quantifies and measures the *similarity* from a database of previously measured observations (Maltamo & Eerikäinen, 2001). In the context of forest inventory, the k-NN method was first introduced in the late 1980's (Kilkki & Päivinen, 1987), applied later for the prediction of standing timber volume by e.g. (Tomppo, 1993) and was later examined in a handful of studies to predict forest stand and individual tree attributes. As stated by e.g. (Haapanen et al., 2004), the k-NN method has been further developed for modelling forest variables and is now operational in Scandinavian countries e.g. in Finnish National Forest Inventory (NFI). It was further integrated as a part of Forest Inventory and Analysis (FIA) program in the Unites States (see (McRoberts & Tomppo, 2007)). The method couples field-based inventory and auxiliary data (e.g. from remote sensing sources) to produce digital layers of measured forest or land use attributes ((Haapanen et al., 2004)). Following the promising results in Scandinavian landscapes achieved by the application of nonparametric methods in prediction/classification of continuous and categorical forest attributes by means of remotely sensed data, the method have recently received a great deal of attention in other parts of the world e.g. in central Europe (Latifi et al., 2011), as the method could be potentially integrated as a cost effective alternative within the regional and national forest inventories.

Apart from the forest inventories conducted in larger scales, the k-NN method has been applied in the context of so-called small-scale forest inventory, in which the accurate and unbiased inventory of small datasets is of major interest. The term 'small area ' commonly denotes a small geographical area, but may also be used to describe a small domain, i.e. a small subpopulation in a large geographical area (Ghosh & Rao, 1994). Sample survey data of a small area or subpopulation can be used to derive reliable estimates of totals and means for large areas or domains. However, the usual direct survey estimators based on the sampled data are often likely to return erroneous outcomes due to the improperly small sample size. This is more crucial in regional forest inventories, where the sample size is typically small since e.g. the overall sample size in a survey is commonly determined to provide specific accuracy at a much higher level of aggregation than that of small areas. In central European forestry context, a small-area domain is of fundamental importance, since the occurrence of multiple forest ownership systems are historically well-established and still frequently occur. This variation bears, in turn, various forest areas which are connected with different requirements in terms of financial and technological resources for forest inventory. In such situations, high expenses are associated with the regular terrestrial surveys (Stoffels, 2009) and the integration of remote sensing and modelling is thus a motivation to reduce the costs. For example, aerial survey with large footprint ALS flights is reported to generate costs to the amount of 1Euro per ha in Germany (Nothdurft et al., 2009). Therefore, an effective strategy of forest inventory should mainly focus on the inventory of such small forest datasets using all the available infrastructures and potentially attainable technological means. The goal should be set to producing reliable (i.e. sufficiently accurate), general (i.e. reproducible) and (approximately) unbiased models of prominent forest attributes which support providing an up-to-date and continuous information database within the bigger framework of periodical state-wide forest inventory system.

However, some issues are crucially required to be taken into consideration, before a remote sensing-supported modelling task of forest attributes can be commenced. These include:

1.3.1 Data combination issues

Remote sensing data provides a valuable source of information to the forest modelling process. The advanced use of 2 and 3D data in both single-tree and area-based approaches of attributes retrieval would offer valuable potentials to characterize the (inherently) 3D structure of the forest stands (particularly vertical structure such as mean or top height). The data combination is specific to the objectives being set within the case study, as well as to the level of details which is required by the analyst. As such, different data including broadband optical (both medium and high spatial resolution), hyperspectral, LiDAR (height as well as intensity), and RADAR data can be combined or fused to reach those goals.

1.3.2 The configuration of models

Depending on what modelling scheme is aimed to be used to retrieve the response forest attributes, a set of parameters are necessary to be set prior to modelling. These parameters can therefore greatly affect issues such as modelling errors and the retrieved values. In case of parametric regression, the underlying distribution of the data, the type of model in use (e.g. Ordinary Least Squares (OLS) or logarithmic models) and model parameters are crucial to be mentioned (see e.g.(Straub & Koch, 2011)). In nonparametric methods, issues like the selection of smoothing parameter for smoothing methods (e.g. (Wood, 2006)), size of neighbourhood for k-NN models, and number of trees per response variable for CART-based methods are necessary to be optimally set. Specifically in terms of k-NN models, the main difference amongst the various approaches is how the distance to the most similar element(s) is measured, which in turn depends on how the *similarity* is quantified within the feature space formed by the multiple predictors. This causes the main difference amongst the diverse distance measures which work based on k-NN approach including the well-known Euclidean and Mahalanobis distances. The neighbourhood size (known also as the number of NNs or k) can be set to any number from 1 to n (the total number of reference units). The single neighbour can, however, contribute to producing more realistic predictions in small datasets, while avoiding major prediction biases in cases where the responses follow skewed (or non-Gaussian) distributions (Hudak et al., 2008). However, one may note that using multiple neighbours would apparently yield more accurate results through averaging values from multiple response units.

1.3.3 Screening the feature space of candidate predictors

When dealing with datasets associated with numerous independent variables, one aim is to reduce the dimensionality of the feature space. Even though heuristic approaches may often be used to deal with highly-correlated variable sets, application of appropriate variable screening methods has recently become an important issue in modelling context. In variable screening, the main objective is to optimize the efficiency of models by achieving a certain performance level with maximum degree of freedom (Latifi et al., 2010). When building models in small scale geographical domains using several (and often strongly inter-correlated) remote sensing metrics, one would most probably come up with the question of how the most relevant information could be extracted from the enormous information content stored in the dataset. This is of major importance when the aim is to build parsimonious models being valid not only across the underlying region of parameterization, but also in further domains which show the (relatively) similar conditions. It also plays a crucial role in k-NN modelling approaches, since the majority of those methods lack an effective built-in scheme for feature space screening. The performances of different deterministic (e.g. forward, backward and stepwise selection methods) and stochastic (e.g. genetic algorithm) have been investigated in various studies available in the literature.

2. Remote sensing for modelling forest structure

2.1 Forest attribute modelling using optical data

Due to the lack of required 3D information for characterisation of vertical structure of forest stands, the pure use of multispectral optical remote sensing for forest structure has severe limitations. (Koch, 2010) addresses this issue and states that those data sources have been mainly employed to differentiate amongst e.g. rough biomass classes which show clear distinctions. For example, Simple linear, multiple, and nonlinear regression models were tested by (Rahman et al., 2007) to classify different levels of forest succession in such as primary and secondary forests, where optical band reflectance and vegetation indices from Enhanced Thematic Mapper (ETM+) data were used as predictors. The use of dummy variables was reported to improve the accuracy of forest attribute estimation by ca. 0.3 of R^2 (best R^2 = 0.542 with 10-13 dummy predictors). In an earlier attempt in central Europe, (Vohland et al., 2007) performed parametric classification for a German test site based on a TM image, where 8 forest types were identified with an overall accuracy of 87.5 %. The Linear Spectral Mixture Analysis (endmember method) was also used to predict stem count, in that the fractions extracted from the spectra were linearly regressed with stem count as response variable. This different approach was also reported to introduce an improved calibration of large-scale forest attribute assessment. Although using parametric approaches, the methodology was (truly) stated to be also helpful in case of using nonparametric approaches. Regarding the observed linear correlations between the response variable of interest (stem count) and spectral indices, this assertion seems to be realistic. The usefulness of Landsat-derived features to model forest attributes (species richness and biodiversity indices) has also been discussed and confirmed by (Mohammadi & Shataee, 2010), in which they reported some positive potentials of multiple regressions (adjusted R^2=0.59 for richness and R^2=0.459 for reciprocal of simpson index) in temperate forests of northern Iran.

Attempts toward establishing correlations amongst regional-scale multispectral remote sensing and forest structural attributes in larger scale dates back to some early attempts in the early 1990's, amongst which e.g. (Iverson et al., 1994) can be highlighted. Their empirical regressions between percent forest cover and Advanced Very High Resolution Radiometer (AVHRR) spectral signatures was used based on Landsat-scale smaller calibration centres. Extrapolating forest cover for much bigger scales (state-scale) using AVHRR data resulted in high correlations (r=0.89 to 0.96) between county cover estimates. Those attempts to produce large-scale maps of forest attributes continued up to some later studies e.g. (Muukkonen & Heiskanen, 2007) and (Päivinen et al., 2009). Whereas regression modelling of AGB using Adavanced Spaceborne Thermal Emission and Radiometer (ASTER) and Moderate Resolution Imaging Spectrometer (MODIS) data was pursued in the former study (relative Root Mean Square Error (RMSE)% = 9.9), the latter used AVHRR pixel values which were applied to be regressed with the standing volume to produce European-scale growing stock maps. (Gebreslasie et al., 2010) can be noted as a very recent effort to parametrically model the forest structure in local scale, in which the visible and shortwave infrared ASTER features (original bands and vegetation indices) were investigated to build stepwise regressions of standing volume, basal area, stem count and tree height in *Eucalyptus* plantations. Whereas the spectral data was acknowledged to be an insufficient material to be solely used for modelling (R^2= 0.51, 0.67, 0.65, and 0.52 for standing volume, basal area, stem count and tree height, respectively), integrating age and site index data as predictors showed to notably enhance the models by 42 %, 20.2%, 16.8%, and 42.2% of R^2. The sole application of multispectral data, regardless of the scale within which the data have been used, seems not to fulfil the practical requirements for accurate regression modelling of forest attributes. Except some very few reports showing highly-correlated spectral indices with stem volume (approximate R^2= 0.95 for multiple linear regression using SPOT and AVHRR data in provincial level reported by (Gonzalez-Alonso et al., 2006)), most of other reports state moderate correlations. However, the majority of the studies have acknowledged the potentials in using such spectral data for regression modelling of forest structural attributes.

In context of nonparametric methods, as documented earlier, the initial introduction of k-NN methods to forestry context commenced in the late 1980's and early 1990 's, as a number of preliminary studies were carried out in the Nordic region. The method was initially in use only based on field measurements (Tomppo, 1991) and was later adapted for prediction of stem volume using spaceborne images. At that time, the most feasible satellite image data included Landsat Thematic Mapper (TM) and SPOT images, from which mainly TM and, to a minor extent, SPOT data were employed (Tomppo, 1993). The reported results have confirmed the suitability of the method based on remote sensing data. The method was further developed through various experiences. The further Finnish experiences with pure optical data include a range of studies in which the k-NN method was attempted to be adapted to practical applications in wood and timber industry. Amongst them, (Tommola et al., 1999) used k-NN method as a tool for wood procurement planning to estimate the characteristics of cutting areas in Finland. They found it to be a useful tool compared to the traditional inventory method. (Tomppo et al., 2001) utilized the approach to estimate/classify growth, main tree species, and forest type by means of multispectral TM data in China. The authors found the method to be helpful in classifying tree types and stand ages, though the stand-level predictions were reported to underestimate the growing stock.

As mentioned above, k-NN estimators include a range of distance-weighting approaches such as conventional distances (Euclidean and Mahalanobis) and Most Similar Neighbour (MSN) method. Due to the importance of those methods in the context of spatial modelling, a brief verbal explanation of those distance metrics seems to be essential: In general, the distance between the target units with a vector of predictor variables to any neighbouring unit having the multi-dimensional vector of predictors can be measured by a distance function, in which the weight matrix of predictors plays a central role to weight the predictors according to their predictive power. Whereas this weight matrix turns to be a multi-dimensional identity matrix (in the Euclidian distance) or the inverse of the covariance matrix of the predictor variables (in the Mahalanobis distance), the MSN inference uses canonical correlation analysis to produce a weighting matrix used to select neighbours from reference units. That is, according to (Crookston et al., 2002), the weight matrix is filled with the linear product of the squared canonical coefficients and their canonical correlation coefficients. The MSN method was described by e.g. (Maltamo & Eerikäinen, 2001) as a closely- related method to the basic k-NN based on Euclidean distance, whereas the main difference is that the coefficients of the variables in the distance function are searched using canonical correlations in MSN. Thus, one should bear in mind that a linear correlation between response(s) and predictor(s) can play a key role in the MSN method. The majority of attempts to construct MSN models of forest structure made use of 3D LiDAR data, either alone or in combination with spectral metrics. Therefore, the literature regarding MSN modelling will further be reviewed in the LiDAR section.

To the best of author's knowledge, Efforts to bring the analytical features of k-NN method to the US NFI system (called Forest Inventory and Analysis, FIA) were accomplished by studies such as (Franco-Lopez et al., 2001) who used the method to simultaneously predict basal area, volume and cover types based on FIA field inventory data and TM features. They truly mentioned a common small-scale problem (i.e. the critical performance of k-NN methods in case of small datasets) and acknowledged that "The key to success is the access to (enough) ground samples to cover all variations in tree size and stand density for each cover type".

(Katila, 2002) integrated TM and forest inventory data to model forest parameters including landuse classes. The results were verified using the Leave-one-out (LOO) cross validation (Efron & Tibshirani, 1993) on the pixel level. The method was assessed to be statistically straightforward comparing to the conventional landcover estimation. (Hölmstrom, 2002) used a set of panchromatic aerial photos and field based information from 255 circular sample plots measured within the boreal forests of Sweden. Stem volume and age were modelled and validated, through which 14 % and 17 % of prediction errors ($RMSE$) for volume and age of the trees were observed, respectively. The k-NN method was thus proposed for stand level applications. However, they highlighted the importance of sufficient and representative reference material and the considerations in selecting the number of neighbours in small datasets as potential drawbacks.

The application of RADAR data in forest assessments has been reported to be associated with some major constraints due to signal saturation (Imhoff, 1995) which can also occur in optical images when the forest canopy is fully closed (Holmström & Fransson, 2003). However, RADAR reflectance has been reported to be linearly related to standwise stem volume (Fransson et al., 2000). Therefore, multispectral data has been combined, though in relatively few experiences, with active data from RADAR platforms for retrieval of forest attributes. For example, (Holmström & Fransson, 2003) tested the fusion of optical SPOT-4 and airborne CARABAS-II VHF Synthetic Aperture RADAR (SAR) datasets to estimate forest variables in Spruce/Pine stands. The single use of each data was compared to the combined use, and the combined data was expectedly assessed to surpass the single one for modelling stem volume and age ($RMSE=37$ $m^3 ha^{-1}$ of combined set compared to $RMSE=50$ $m^3 ha^{-1}$ of the best single-data models). The relationship between the reference target units was reported to be "substantially strengthened" when using the two data sources in combination. Later on, (Thessler et al., 2008) investigated the joint application of multispectral and RADAR data in an alternative workflow to the one explained above, in that they applied TM-derived features combined with predictors extracted from the Digital Elevation Model (DEM) of a shuttle RADAR data to classify the tropical forest types in Costa Rica. Some cover type classes were consequently merged to aggregate the classes and improve the results, which led to the overall accuracy of 91 % from the segmented image data based on k-NN classification. (Treuhaft et al., 2003) combined C-band SAR interferometry with Leaf Area Index (LAI) extracted from hyperspectral data to estimate AGB. They introduced their resulted 'forest canopy leaf area density' to be a representative for AGB of forest.

Though the conventional k-NN models of stand-scale forest attributes have been positively supported in the studies like those mentioned above, some other studies e.g. (Finley et al., 2003) acknowledge that the analysts may face the challenge of compromising between increased mapping efficiency and a loss of information accuracy. This is particularly the case when dealing with the question of selecting the optimal number of neighbours (also known as k). Different neighbourhood sizes have been studies in several works ((Franco-Lopez et al., 2001), (Haapanen et al., 2004),(Holmström & Fransson, 2003), (Packalén & Maltamo, 2006), (Packalén & Maltamo, 2007), (Finley & McRoberts, 2008) and (Vauhkonen et al., 2010)), in some of which the optimum number of k were discussed ((Franco-Lopez et al., 2001), (Haapanen et al., 2004), (Finley & McRoberts, 2008)). Whereas the above- mentioned studies reported an improved accuracy of k-NN predictions along with the increment of k (up to a limited number varying amongst the studies), some acknowledge that increasing k leads to a stronger shift of the predictions towards the sample mean which could cause serious biases, particularly in cases where the distribution of observations is skewed ((Hudak et al., 2008), (Latifi et al., 2010)). However, the choice of neighbourhood size is an arbitrary issue in which the expertise of the analyst (e.g. the prior knowledge on the properties and variance of the population) plays a functional role. By using multiple k for imputation, the majority of studies carried out within the framework of FIA program in US (characterized by a cluster sampling design using 4 subplots in each cluster) have shown to yield relatively high accuracies. The study of (Haapanen et al., 2004) can be exemplified, in which three classes of forest, non-forest and water were classified by a conventional k-NN approach (Euclidean distance) and ETM+ features as predictors. They increased the neighbourhood size up to 10 neighbours, which caused an enhancement of overall accuracy up to the use of 4th neighbour, a sudden drop, and a consequent improvement up to $k=8$. The Majority of other studies in this realm have reported the improvement of accuracy along with increment in the neighbourhood size. Some studies noticed that the selection of other parameters such as weighting distances also depends on the choice of image dates and other associated data ((Franco-Lopez et al., 2001), (Finley & McRoberts, 2008)). (Mäkelä & Pekkarinen, 2004) made a relatively preliminary effort to use field data of stand volume from an inventoried area to make predictions in a neighbouring region which was considered to suffer from lack of field data. However, their poor accuracy yielded from the estimation led them to assess the method as an inappropriate one for stand level predictions. Yet, some of their best volume estimates were reported to be useful for the stands where no (or few) field information is available. In a study conducted in a central Europe, (Stümer, 2004) developed a k-NN application in Germany to model and map basal area (i.e. metric data) and deadwood (i.e. categorical data) using TM, hyperspectral, and field datasets as predictors. The best results showed the RMSE between 35 % and 67 % (for TM data) and 65 % and 67 % (for hyperspectral data). As for the deadwood, the accuracy ranged between 60 % and 73 % (for TM) and 60 % and 63 % (for hyperspectral). The two data sets were separately assessed, in which no combinations were tested.

Using various configurations of k-NN methods, (LeMay & Temesgen, 2005) compared some combinations (e.g. varying number of neighbours) to predict basal area and standing volume in Canadian forests. They reported MSN method (even in a single-neighbour setting) as the most accurate approach compared to the Euclidean distance models based on 3 neighbours. In a relatively similar study in Bosnian forests in Europe, (Cabaravdic, 2007) also achieved relatively accurate k-NN estimates of growing stock using TM-extracted features and a broad range of field survey information. In terms of the configuration, $k=5$ and Mahalanobis distance were assessed to be optimal for growing stock models. (Kutzer, 2008) tested the selected bands in visible and infrared domain of multispectral ASTER image together with a set of terrestrial data to differentiate the landuse types and the Non Wood Forest Products in Ghana. The results were assessed, though with some exceptions, to be promising for application as a practical forest monitoring tool within the study area.

The majority of forest-related studies using k-NN method have been conducted with the aim of modelling continuous attributes of forest structure, whereas little attention has been paid to predicting categorical forest variables such as site quality or vegetation type. One of the few attempts to introduce such new potentials to the remote sensing society was carried out by (Tomppo et al., 2009), in which TM-derived spectral features were used to predict site fertility, species dominance and coniferous/deciduous dominance as categorical responses across selected test sites in Finland and Italy. Despite the moderate accuracy obtained out of the sole analysis of spectral data (e.g. max. Kappa statistics of approximately 0.65 and relatively higher Kappa values of species dominance compared to soil fertility), this study highlighted the importance of how an efficient strategy for feature space screening can contribute to reducing the prediction errors in k-NN models. Whereas the majority of pearlier studies used deterministic approaches (e.g. stepwise methods) to prune the candidate predictors, this study (which followed an earlier attempt by (Tomppo & Halme, 2004) used an evolutionary Genetic Algorithm (GA) to screen the feature space which reduced the modelling errors in slight rates. The idea of using GA was further applied for a number of LiDAR-supported forest modelling studies by e.g. (Latifi et al., 2010) and (Latifi et al., 2011).

2.2 LiDAR-based models of forest structural attributes

Height information from airborne laser scanner data has been validated to provide the most accurate input data related to the topography of land surface as well as to the structure of forested areas. Whereas (Lim et al., 2003), (Hyyppä et al., 2008) and (Koch, 2010) provide comprehensive reviews on the background and history of LiDAR data application in forest inventories, this section focuses on the methodological background concerning pure LiDAR-based models of forest structure.

LiDAR instruments include three main categories of profiling, discrete return, and waveform devices. Profiling devices record one return at low densities along a narrow swath (Evans et al., 2009) and were mainly used in the earlier studies such as (Nelson et al., 1988). Later, discrete-return (Pulse form) laser scanners enabled to use LiDAR in remote sensing where scanning over large areas was needed (Næsset, 2004). Such devices collect multiple returns (often three to five returns) based on intensity of the emitted laser energy from the earth surface. In terms of waveform data, the devices digitize the total amount of emitted energy in intervals and therefore are able to characterize the distribution of emitted laser from the objects. Although small footprint waveform sensors are most commonly available, they are reported to be computationally intensive and thus associated with restrictions when used in fine-scale (i.e. high resolution) environmental applications (Evans et al., 2009). They provide data featuring high point densities and enable one to broader representation of the surface and forest canopy. The importance of using pulse form data for studies concerning forest structure is already stated in the relevant literature e.g. (Sexton et al., 2009).

LiDAR data can be used in two main approaches to retrieve forest structural attributes. In "area-based methods", the statistical metrics and other nonphysical distribution-related features of LiDAR height measurements are extracted either from the laser point clouds or from a rasterized representation of laser hits. They are then used to predict forest attributes e.g. mean tree height, mean DBH, basal area, volume and AGB at an area-level such as the plot or stand level (Yu et al., 2010). This method enables one to retrieve canopy height information by means of a relatively coarse resolution LiDAR data e.g. satellite or airborne data featuring <5 measurements per m^2 e.g. (Korhonen et al., 2008), (Jochem et al., 2011), though data with higher point density can also be used to derive the metrics at an aggregated level (e.g. (Maltamo, Eerikäinen, Packalén & Hyyppä, 2006) (Heurich & Thoma, 2008), (Straub et al., 2009) and (Latifi et al., 2010)). A key to success in area-based methods, when the metrics are extracted from a rasterized form of LiDAR data such as normalized Digital Surface Model (nDSM), has been stated to be the quality of extracted Digital Terrain Model (DTM) and Digital Surface Model (DSM) (Hyyppä et al., 2008).

The focus in the so called "Single tree-based methods" is on the recognition of individual trees. Here, the tree attributes e.g. tree height, crown dimensions and species information are measured. The measured attributes can further be applied to retrieve other attributes such as DBH, standing volume and AGB by means of various modelling approaches (Yu et al., 2010). The retrieved attributes are either presented as single-tree attributes or can be aggregated into a higher level e.g. stand or sample plot level.

In some earlier studies, one of the main goals in applying 3D data was to facilitate an accurate estimation of stand height, in which correlating the laser-derived height information to those measured in the field was of major interest. This often yielded notably promising results which strongly supported the accuracy of LiDAR instruments for precise height measurements. For example, (Maltamo, Hyyppä & Malinen, 2006) used airborne laser data to retrieve crown height information i.e. basal area, mean diameter and height at both tree and plot levels using linear regression methods in Finland. The results indicated the superiority of LiDAR-based attributes over the field-based ones in area-level, though a contrasting result was reported in single-tree level. Better result was hypothesized to be achieved when data with higher point density would be obtained with large swaths. The roughly similar result was later reported by (Maltamo, Eerikäinen, Packalén & Hyyppä, 2006), in which the plot-level stem volume estimates calculated from field assessments were reported to be less accurate than the methods in which volume had been predicted by LiDAR measures.(Maltamo et al., 2010) further studied different methods including regression models to retrieve crown height information. Regardless of the differences amongst the methods, they all yielded RMSEs between 1.0 and 1.5 m in predicting crown height.

Application of laser scanner data to enhance volume and AGB models dates back to some preliminary experiments in 1980's e.g. (MacLean & Krabill, 1986), (Nelson et al., 1988) which demonstrated the usefulness of LiDAR-extracted canopy profiles to improve stem volume and AGB estimates (e.g.R^2=0.72 to 0.92 achieved in regression analysis by (MacLean & Krabill, 1986)). In the recent years, except some cases, the investigations on further developments in the retrieval of model-derived volume and AGB attributes has considerably grown. (Heurich & Thoma, 2008) built linear models to predict plot-level stem volume, height, and stem count in Bavarian National Park, where they reported $RMSE\%$ =5, 10 and 60 for LiDAR-estimated height, volume and stem count, respectively. The forest areas were stratified into three main deciduous, coniferous, and mixed strata. Despite achieving relatively accurate results in their models, they acknowledged that factors such as occurrence of deadwoods and complexities in forest structure constrain the achievement of better results. As stated earlier, derivation of model-based estimates of stem volume (in different assortments) have recently formed a major field of research in LiDAR-related studies. The Sawlogs can be exemplified as vital timber assortments in Nordic forest utilization context. Therefore, the accurate estimation of their volume can lead to an added value in forest management. (Korhonen et al., 2008) studied this by using parametric models, in that they used LiDAR canopy height metrics i.e. percentiles to make linear models of sawlog volume, which yielded relatively favourable accuracies ($RMSE\%$=9.1 and 18 for theoretical and factual volumes). In other examples, regression modelling of individual trees using the multi-return, pulse-form LiDAR metrics has been reported to be accurate for standing volume (R^2=0.77) (Dalponte et al., 2009) as well as for AGB (Max. R^2=0.71) (Jochem et al., 2011).

In terms of the type of metrics extracted from laser scanner data, one important issue cannot be neglected: In addition to height metrics, the LiDAR intensity data is reported to contain some information in infrared domain which may potentially share some values to the modelling of forest attributes e.g. (Boyd & Hill, 2007), especially when dealing with species-specific models (Koch, 2010). Regardless of some exceptions e.g. (Vauhkonen et al., 2010),(Latifi et al., 2010), most of the pure LiDAR-based models of forest attributes solely made use of height metrics as input variables for modelling.

Using nonparametric methods greatly contributed to the studies aiming at retrieval of forest attributes by means of LiDAR metrics. Those methods have been applied in various scales, using numerous metrics, and combined, in some cases, with additional methods for screening the high-dimensional feature space or for estimating the prediction variance. (Falkowski et al., 2010) evaluated k-NN imputation models to predict individual tree-level height, diameter at breast height, and species in northeastern Oregon in USA. Topographic variables were added to LiDAR-extracted height percentiles and other descriptive statistics to accomplish the task. Whereas 5 and 16 m^3ha^1 of $RMSE$ were achieved for basal area and volume estimates, occurrence of small trees or the dense understory showed to be the main source of prediction errors. Similarly, promising results have been reported by e.g.(Nothdurft et al., 2009) in central Europe for area-based models of stem volume using LiDAR height metrics (approximately 20 % of $RMSE$ for MSN models of stem volume in Germany).

(Hudak et al., 2008) compared different imputation methods to impute a range of forest inventory attributes in plot level using height metrics from LiDAR data and additional topographical attributes in Idaho, USA. They found the Random Forest (RF) to be superior to other imputation methods such as MSN, Euclidean distance and Mahalanobis distance. They used the selected RF outputs for final wall-to-wall mapping of forest structural attributes at pixel level. The dominance of RF model was further confirmed by studies such as (Latifi et al., 2010) and (Breidenbach, Nothdurft & Kändler, 2010) and led to a wider application of RF as a leading nonparametric method in combination with LiDAR metrics e.g. (Yu et al., 2011). The RF method (Breiman, 2001) works based on ensembles of CARTs for resampled predictor variable sets. It starts with evolving bootstrap samples from the original data. It then grows, for each bootstrap sample, an unpruned regression tree. The best splits are chosen from the randomly sampled variables at each node or the trees. The new predictions are then made by aggregating the predictions of the total number of trees. That is, the mode votes (the most frequent values) from the total trees will be the predicted value of the respective variable ((Liaw & Wiener, 2002), (Latifi et al., 2011)). Though the former studies e.g. (Hudak et al., 2008) and (Vauhkonen et al., 2010) have shown that the RF approach generally surpasses other imputation methods including MSN, (Breidenbach, Nothdurft & Kändler, 2010) reported an approximately similar performance of RF and MSN, as their study yielded e.g. the $RMSE$ of 32.41 % (for MSN) and 32.81 % (for RF) when predicting the total standing timber volume by averaging k=8.

In addition to those stated above, the nonparametric methods were also tested to predict further structural characteristics of forest stands e.g. diameter distributions by the sole use of laser scanner data (e.g. (Maltamo et al., 2009)), yielding some potentials towards further application of 3D topographic remote sensing for forest monitoring.

2.3 Combining LiDAR and optical data for modelling

As explained earlier, the application of ALS-extracted metrics (height and intensity features) has been validated as a being helpful and thus required for most practices regarding forest inventory. This is because the data has previously been proved to be potentially applicable in several environmental and natural resource planning tasks, particularly where the vertical structure of the respective phenomena is dealt with. Nevertheless, the use of multi-sensorial data may enable one to make use of advanced methods of data analysis and thus overcome some problems faced by using single datasets (Koch, 2010). The use of multispectral data can contribute to the analysis of vegetation cover by adding spectral information from visible and infrared domains. In this way, the information required for species-specific tasks will be provided by the spectral data, while the LiDAR data contributes an enormous amount of information in terms of 3D structural attributes (see e.g. (Packalén & Maltamo, 2007), (Heinzel et al., 2008), (Straub et al., 2009)).

When combining spectral and LiDAR data, the parametric models have been quite rarely used for predicting forest attributes. In contrast, relatively more studies were carried out using combined data made use of nonparametric methods (especially MSN and RF), probably as the models are generally assumed as rather 'distribution-free methods' which can potentially be applied regardless of the underlying distribution of the population. A further reason could be the ability of more advanced methods such as MSN and RF to handle high-dimensional feature spaces. However, examples of the joint use of spectral and laser scanner data for parametric modelling can be e.g. (Fransson et al., 2004) and (Hudak et al., 2006), in both of which the magnitude of candidate predictors were notably less than those making use of nonparametric methods. (Fransson et al., 2004) built regression models to predict stem volume using SPOT5 data aided by TopEye laser scanner data in Swedish coniferous landscapes. The SPOT5 data was used to develop features including multi-spectral bands, ditto squared, and the band ratios. LiDAR- derived features included height and forest density measures at stand level. The single as well as combined datasets were tested, from which the combined use of laser height data with the spectral features surpassed the individual use of the datasets. Later on,(Hudak et al., 2006) linearly regressed basal area and tree density on 26 predictors derived from height/intensity of LiDAR and Advanced Land Imager (ALI) multispectral data. They found laser height (to a higher extent) added by laser intensity metrics as most relevant predictors of both responses (The LiDAR-dominated models explained around 90 % of variance for both response variables).

In terms of applying conventional distance-based k-NN methods, (McInerney et al., 2010) can be referred who combined airborne laser scanner and spaceborne Indian Remote Sensing (IRS) multispectral data to model stand canopy height using k-NN method. They apparently reported laser height data as the major means of canopy height retrieval, and achieved a relative $RMSE$ between 28 and 31 %. (Maltamo, Malinen, Packalén, Suvanto & Kangas, 2006) applied a k-MSN (MSN using multiple k) method to combine the LiDAR data with aerial images and terrestrial stand information in Finland. The laser-based models were reported to outperform aerial photography in stand volume estimation, and the combination improved the models at plot and stand levels. (Wallerman & Holmgren, 2007) have also highlighted the combined application of predictive features derived from optical (SPOT) and laser (TopEye) data, according to which the combined dataset yielded the mean standing volume and stem density models with $RMSE = 20\%$ and $RMSE = 22\%$, respectively. Combining satellite-based (TM) spectral features with laser metrics was also carried out by (Latifi et al., 2010) who reported that TM-extracted metrics can be used as alternatives to those derived from aerial photography for area-based models. Using k-MSN approach, (Packalén & Maltamo, 2006) conducted a survey to achieve species-specific stand information using sets of aerial photography and ALS data. The procedure consisted of two methods including 1) simultaneous k-MSN estimation and 2) a two- phase prediction (prediction of the responses using regression analysis of ALS data and then allocation of the variables using a fuzzy classification approach). The k-MSN achieved better results than the fuzzy classifications. Although the study still proposed some further developments of the predictor variables from both datasets, the results were assessed satisfactory in cases of Norway spruce (*Picea abies* L.) and Scots pine (*Pinus sylvestris* L.). Soon after, (Packalén & Maltamo, 2007) made stand level models of volume and height using the similar dataset as before. A set of Haralick textural features(Haralick, 1979) from the optical data were additionally combined with the calculated ALS height features to produce predictive models. Accuracy of the predicted responses was finally found to be comparable to stand-level field assessments, though the attributes of conifers were estimated more accurately than those from the deciduous stands. In a further study by those authors, (Packalén & Maltamo, 2008) made use of the similar data to develop k-MSN models of diameter distribution by tree species. Based on the results of growing stock estimation in the previous research work(s), two approaches were compared including 1) field-based modelling using the Weibull distribution and 2) k-MSN prediction, in which the latter was assessed to outperform the former method. Nevertheless, the need to have more comprehensive reference field data (i.e. a common small-scale problem) to cover the spectral variations of the remote sensing data was highlighted as a major concern which supports those already acknowledged by precedent studies. (Nothdurft et al., 2009) represents an attempt towards solving this, in which bootstrap-simulated prediction errors of MSN inferences of volume based on sole use of LiDAR height metrics were smaller than those of design-based sampling.

Few studies e.g. (Straub et al., 2010) and (Latifi et al., 2011) compared parametric and nonparametric methods for forest attribute estimation in presence of both LiDAR and multispectral datasets. Whereas the former study compared Ordinary Least Squares (OLS) regression and a yield table-estimated stem volume with that from Euclidean distance-based k-NN method, the latter made a comparison between RF and OLS outputs. Nevertheless, both studies made relatively similar conclusions, in that they stated that using nonparametric methods cannot b expected to remarkably contribute to the improvement of forest attribute estimates. Besides, it supports (Yu et al., 2011)who also tested pure LiDAR metrics and achieved a similar performance of RF and OLS in a single tree scale. The rationale behind this is that non-parametric imputations do not share the same mix of error components as regression predictions. Imputation errors are often greater than regression errors because the errors do not result from a least-squares minimisation, but from selection of a most similar element in a pool of neighbouring observations (Stage & Crookston, 2007). However, K-NN methods (especially in single- neighbour setting) yield predictions with similar variance structure to that of the observations (Moeur & Stage, 1995), and are thus advantageous over the higher accuracies achievable by the use of OLS (Hudak et al., 2008).

The selection of proper predictor variables for a k-NN model (i.e. an absent element of conventional k-NN approaches) is a time-consuming task which and needs to be automated. (Packalén & Maltamo, 2007) used an iterative cost- minimizing variable selection algorithm which aimed at minimizing the weighted average of the relative $RMSE$. In contrast, studies like (Hudak et al., 2008) and (Straub et al., 2009)applied stepwise selection methods, where the former study based its stepwise iteration on the *Gini* index of variable importance used by (Breiman, 2001) as a built-in feature in RF. As such, other variable screening methods such as parametric univariate correlation analysis (Breidenbach, Næsset, Lien, Gobakken & Solberg, 2010), Built-in schemes of RF such as stepwise iterative method (Vauhkonen et al., 2010) and forward selection (Breidenbach, Nothdurft & Kändler, 2010)were also used to complete this task in the recent literature. Each of those screening methods has been reported to be satisfying in terms of reducing the dimensionality of the feature space, though no rationale (e.g. comparison to other methods) has been presented. (Latifi et al., 2010) used a GA on categorised response variables to optimise the high-dimensional feature space formed by numerous correlated predictors. Even though this GA prototype was evaluated to efficiently reduce the relative RMSE of standing volume and AGB compared to the stepwise selection of predictors, the method was reported to produce unstable subsets attributed to strong correlations amongst the predictors. By using a Tau-squared index on continuous responses, GA was later shown to yield stable parsimonious variable subsets (Latifi et al., 2011). GA is a search algorithm which works via numerous solutions and generations and thus explores the entire possible combinations of candidate predictor variables. It provides the consequent NN models with the optimum range of refined, pre-processed feature space formed of relevant (and uncorrelated) remote sensing descriptors and is shown to be able to be adjusted to the k-NN modelling approaches (e.g. (Tomppo & Halme, 2004)). In this context, fitness functions to optimise continuous responses are preferable for regression scenarios. Those functions can even be linear as long as no highly non-linear trend/prediction is observed in the entire underlying dataset.

In a review by (Koch, 2010), the importance of combined use of laser and optical data for such purposes was highlighted. She stated that combining the altimetric height information with physical values derived from laser intensity is appropriate for modelling forest structure. As 3D data has already been shown to be plausible for AGB modelling, and due to the expected future technical innovations of those data for biomass assessments, it is assumed that it will further play a prominent role in major forest monitoring tasks e.g. those related to AGB modelling.

3. Conclusion

Amongst the available active/passive remote sensing instruments, information derived from laser scanner (especially the height information) is definitely of major importance for studies regarding forest structure. According to (Koch, 2010), the significance of using LiDAR data for biomass assessment has been confirmed by variety of investigations which repeatedly showed comparatively higher performance of those data. However, the use of LiDAR intensity data is still limited. The intensity data has been shown to be able to add useful complementary information to LiDAR height data for forest attribute modelling (e.g. (Hudak et al., 2006)). Yet, a direct physical connection between those intensity metrics and forest structure still cannot be drawn. The reason for this complication is stated to be the dependency of intensity on a range of factors affecting reflected laser data including range, incidence angle, bidirectional reflectance function effects, and transmission of atmosphere (Hyyppä et al., 2008).

Apart from few exceptional studies which reported the incapability of spectral data for explaining the variation beyond the variation that could be explained by laser metrics (Hudak et al., 2008), adding spectral information to pure LiDAR-based models has been confirmed to be useful, as they provide continuous information over long time series and are spectrally sensitive for differentiating tree species. The ability of multispectral data, even in regional-scale spatial resolution such as Landsat images, has been constantly approved to bear practical values when combined with laser scanner data ((Fransson et al., 2004), (McInerney et al., 2010)) and even as an alternative to aerial photography for area-based applications (Latifi et al., 2010). Furthermore, image spectroscopy data showed positive potentials for forest modelling ((Foster et al., 2002), (Schlerf et al., 2005)) and could potentially complement LiDAR-based models. However, one should bear in mind that the experimental results of surveys is by no means an eventual justification for the small- scale end users to take the acquisition of (relatively) expensive airborne hyperspectral data for granted.

In terms of various modelling methods used, both parametric and nonparametric modelling categories were frequently employed to describe the forest structural attributes. However, the latter approaches received more attention during the recent years to be run for high dimensional predictor datasets as well as for simultaneous predictions. The k-NN methods (especially MSN and RF) have been successfully coupled with LiDAR information and thus caused a rapid increase in the number of research projects during recent years. As it was shown here, much work has been done on area-based methods e.g. stand and plot levels, whereas single-tree approaches still lack some research, mainly due to high computational requirements and the need for high resolution data.

In terms of handling predictor feature space induced by remote sensing features, some examples were previously referred. Whereas studies such as (Breidenbach, Nothdurft & Kändler, 2010)made the general necessity of variable screening in k-NN context questionable, some other studies acknowledge the requirement to selecting an effective strategy of pruning of predictor dataset (e.g. (Hudak et al., 2008), (Latifi et al., 2010)) and showed some decisive influences on the outcomes of the forest attribute models. The proper pruning of predictor feature space has been proved to help producing robust models (Latifi & Koch, 2011). Reducing the sensitivity of models has been also shown to greatly contribute to increasing the robustness of the models. Using resampling methods e.g. bootstrapping to reproduce the underlying population (e.g. (Nothdurft et al., 2009),(Breidenbach, Nothdurft & Kändler, 2010), and (Latifi et al., 2011) increases the potential and robustness of applying nonparametric models in small-scale forest inventory, where the shortage of reference data for validating the models is a major constraint. Robust models enable the analyst to apply them under other natural growing conditions except of the underlying test site, and can thus open up new operational applications for the yielded models (e.g. (Koch, 2010)).

Along with the rapid advancements in launching the active/passive remote sensing instruments, the general access to high resolution products (particularly to laser scanner data) at reasonable costs is increasing. Therefore, the efforts towards thorough description of tree and forest stand structure are currently following a boosting trend all over the world. However, it is necessary to emphasize, again, that much care should be taken in terms of producing valid and robust results, as well as to get the best out of the available data and modelling facilities. Whereas the rapid and accurate modelling of standing volume, biomass and tree density is still important, some remaining open areas of research still require further research. These include, for example, efforts towards advanced classification tasks (especially on single-tree level or in complicated mixed stands), modelling understory and regenerations (e.g. important for intermediate silvicultural practices), and modelling rare and ecologically-valuable populations.

4. References

Acker, S., Sabin, T., Ganio, L. & McKee, W. (1998). Development of old-growth structure and timber volume growth trends in maturing douglas-fir stands, *Forest Ecology and Management* 104: 265– 280.

BMU (2009). National biomass action plan for germany, *Technical report*, Bundesministerium für Umwelt, Naturschutz und Reaktorsicherheit (BMU), 11055 Berlin, Germany.

Boyd, D. S. & Hill, R. A. (2007). Validation of airborne lidar intensity values from a forested landscape using hymap data: preliminary analysis, *Proceedings of the ISPRS Workshop ŚLaser Scanning 2007 and SilviLaser 2007Ś Part 3 / W52, Espoo-Finland*.

Breidenbach, J., Kublin, E., McGaughey, R., Andersen, H. & Reutebuch, S. (2008). Mixed-effects models for estimating stand volume by means of small footprint airborne laser scanner data, *Photogrammetric Journal of Finland* 21(1): 4–15.

Breidenbach, J., Nothdurft, A. & Kändler, G. (2010). Comparison of nearest neighbour approaches for small area estimation of tree species-specific forest inventory attributes in central europe using airborne laser scanner data, *European Journal of Forest Research* 129(5): 833–846.

Breidenbach, J., Næsset, E., Lien, V., Gobakken, T. & Solberg, S. (2010). Prediction of species specific forest inventory attributes using a nonparametric semi-individual tree crown approach based on fused airborne laser scanning and multispectral data, *Remote Sensing of Environment* 114: 911–924.

Breiman, L. (2001). Random forests, *Machine Learning* 45: 5–32.

Cabaravdic, A, A. (2007). *Efficient Estimation of Forest Attributes with k NN*, PhD thesis, Faculty of Forest and Environmental Studies, University of Freiburg.

Crookston, N. L., Moeur, M. & Renner, D. (2002). *Users guide to the most similar neighbor imputation program version 2.00*, RMRS-GTR-96.Ogden, UT: USDA Forest Service Rocky Mountain Research Station.

Dalponte, M., Coops, N. C., Bruzzone, L. & Gianelle, D. (2009). Analysis on the use of multiple returns lidar data for the estimation of tree stems volume, *IEEE Journal of Selected Topics in Applied Earth Observations and Remote Sensing* 2(4): 310–318.

Davey, S. (1984). *Possums and Gliders*, Australian Mammal Society, Sydney, chapter Habitat preferences of arboreal marsupials within a coastal forest in southern New South Wales, pp. 509– 516.

Efron, B. & Tibshirani, R. J. (1993). *An introduction to the bootstrap*, New York: Chapman & Hall.

Evans, J. S., Hudak, A. T., Faux, R. & Smith, M. (2009). Discrete return lidar in natural resources: Recommendations for project planning, data processing, and deliverables, *Remote Sensing* 1: 776–794.

Falkowski, M. J., Hudak, A. J., Crookston, N. L., Gessler, P. E., Uebler, E. H. & Smith, A. M. S. (2010). Landscape-scale parameterization of a tree-level forest growth model: a k-nearest neighbor imputation approach incorporating lidar data, *Canadian Journal of Forest Research* 40: 184–199.

Finley, A., Ek, A. R., Bai, Y. & Bauer, M. E. K. (2003). Nearest neighbour estimation of forest attributes: Improving mapping efficiency, *Proceedings of the fifth Annual Forest Inventory and Analysis Symposium*, pp. 61–68.

Finley, A. O. & McRoberts, R. E. (2008). Efficient k-nearest neighbour searches for multi-source forest attribute mapping, *Remote Sensing of Environment* 112: 2203–2211.

Foster, J., Kingdon, C. & Townsend, P. (2002). Predicting tropical forest carbon from eo-1 hyperspectral imagery in noel kempff mercado national park, bolivia, . *IEEE International Geoscience and Remote Sensing Symposium, 2002. IGARSS '02. Vol. 6,*, pp. 3108–3110.

Franco-Lopez, H., Ek, A. R. & Bauer, M. E. (2001). Estimation and mapping of forest stand density, volume, and cover type using the k-nearest neighbours method, *Remote Sensing of Environment* 77: 251?274.

Franklin, J. (1986). Thematic mapper analysis of coniferous forest structure and composition, *International Journal of Remote Sensing* 7: 1287 – 1301.

Fransson, J., Gustavsson, A., Ulander, L. & Walter, F. t. (2000). Towards an operational use of vhf sar data for forest mapping and forest management, *in* T. Stein (ed.), *Proceedigs of IGARSS 2000*, IEEE, Piscataway, NJ., p. 399Ũ401.

Fransson, J., Magnusson, M. & Holmgren, J. (2004). Estimation of forest stem volume using optical spot-5 satellite and laser data in combination, *Proceedings of IGARSS 2004*, pp. 2318–2322.

Gebreslasie, M. T., Ahmed, F, B. & Van Aardt, J. (2010). Predicting forest structural attributes using ancillary data and aster satellite data, *International Journal of Applied Earth Observation and Geoinformation* 125: 523–526.

Ghosh, M. & Rao, J, N. K. (1994). Small area estimation: An appraisal, *Statistical Science* 9(1): 55–76.

Gonzalez-Alonso, F., Marino-De-Miguel, S., Roldan-Zamarron, A., Garcia-Gigorro, S. & Cuevas, J. M. (2006). Forest biomass estimation through ndvi composites. the role of remotely sensed data to assess spanish forests as carbon sinks, *International Journal of Remote Sensing* 27(24): 5409–5415.

Guo, X. J. A. (2005). *climate- sensitive analysis of lodgepole pine site index in alberta*, Master's thesis, Dept. of Mathematics and Statistics. Concordia University, Montreal-Canada.

Haapanen, R., Ek, A. R., Bauer, M. E. & Finley, A. O. (2004). Delineation of forest/nonforest land use classes using nearest neighbour methods, *Remote Sensing of Environment* 89: 265–271.

Haralick, R. M. (1979). Statistical and structural approaches to texture. proceedings, *Proceedings of the IEEE*, Vol. 67(5), pp. 786–804.

Heinzel, J., Weinacker, H. & Koch, B. (2008). Full automatic detection of tree species based on delineated single tree crowns - a data fusion approach for airborne laser scanning data and aerial photographs, *Proceedings of SilviLaser 2008*, Edinburgh, UK, pp. 76–85.

Heurich, M. & Thoma, F. (2008). Estimation of forestry stand parameters using laser scanning data in temperate, structurally rich natural european beech (fagus sylvatica) and norway spruce (picea abies) forests, *Forestry* 81(5): 645–661.

Hölmstrom, H. (2002). Estimation of single tree characteristics using the knn method and plotwise aerial photograph interpretations, *Forest Ecology and Management* 167: 303–314.

Holmström, H. & Fransson, E. S. (2003). Combining remotely sensed optical and radar data in knn estimation of forest variables, *Forest Science* 49(3): 409–418.

Härdle, W. (1990). *Econometric society monographs*, Econometric society monographs, Cambridge University Press, chapter Applied nonparametric regression.

Härdle, W., Müller, M., Sperlich, S. & Werwatz, A. (2004). *Non-parametric and semiparametric models*, Springer, New York.

Hudak, A., Crookston, N., Evans, J., Hall, D. & Falkowski, M. (2008). Nearest neighbour imputation of species-level, plot-scale forest structure attributes from lidar data, *Remote Sensing of Environment* 112: 2232–2245.

Hudak, A. T., Crookston, N. L., Evans, J. S., Falkowski, M. J., Smith, A. M. S. & Gessler, P. (2006). Regression modeling and mapping of coniferous forest basal area and tree density from discrete- return lidar and multispectral satellite data, *Canadian Journal of Remote Sensing* 32: 126–138.

Hyyppä, J., Hyyppä, H., Leckie, D., Gougon, F., Yu, X. & Maltamo, M. (2008). Review of methods of small-footprint airborne laser scanning for extracting forest inventory data in boreal forests, *International Journal of Remote Sensing* 29(5): 1339–1336.

Imhoff, M. (1995). Radar backscatter and biomass saturation: ramifications for global biomass inventory, *IEEE Transactions on Geoscience and Remote Sensing* 33(2): 510–518.

Iverson, L. R., Cook, E. A. & Graham, R. L. (1994). Regional forest cover estimation via remote sensing: the calibration center concept, *Landscape Ecology* 9(3): 159–174.

Jochem, A., Hollaus, M., Rutzinger, M. & Höfle, B. (2011). Estimation of aboveground biomass in alpine forests: A semi-empirical approach considering canopy transparency derived from airborne lidar data, *Sensors* 11: 278–295.

Katila, M., T. E. (2002). Stratification by ancillary data in multisource forest inventories employing k-nearest neighbour estimation, *Canadian Journal of Forest Research* 32: 1548–1561.

Kilkki, P. & Päivinen, R. (1987). Reference sample plots to combine field measurements and satellite data in forest inventory, *Remote Sensing-Aided Forest Inventory. Proceedings of Seminars organised by SNS, 10-12 Dec. 1986, Hyytiälä, Finland. Research Notes No 19. Department of Forest Mensuration and Management, University of Helsinki.*

Kimmins, J. (1996). *Forest ecology*, Macmillan Inc., New York.

Koch, B. (2010). Status and future of laser scanning, synthetic aperture radar and hyperspectral remote sensing data for forest biomass assessment, *ISPRS Journal of Photogrammetry and Remote Sensing* 65: 581–590.

Koch, B., Straub, C., Dees, M., Wang, Y. & Weinacker, H. (2009). Airborne laser data for stand delineation and information extraction, *International Journal of Remote Sensing* 30(4): 935–963.

Korhonen, L., Peuhkurinen, J., Malinen, J., Suvanto, A., Malatamo, M., Packalén, P. & Kangas, J. (2008). The use of airborne laser scanning to estimate sawlog volumes, *Forestry* 81(4): 499–510.

Kutzer, C. (2008). *Potential of the kNN Method for Estimation and Monitoring off-Reserve Forest Resources in Ghana*, PhD thesis, Faculty of Forest and Environmental Studies, University of Freiburg.

Latifi, H. & Koch, B. (2011). Generalized spatial models of forest structure using airborne multispectral and laser scanner data, *Proceedings of ISPRS Workshop: High resolution earth imaging for geospatial information,,* Vol. XXXVIII-4/W19. of *International Archives of the Photogrammetry, Remote sensing and Spatial Information Sciences,,* Hannover, Germany.

Latifi, H., Nothdurft, A. & Koch, B. (2010). Non-parametric prediction and mapping of standing timber volume and biomass in a temperate forest: application of multiple optical/lidar Üderived predictors, *Forestry* 83(4): 395–407.

Latifi, H., Nothdurft, A., Straub, C. & Koch, B. (2011). Modelling stratified forest attributes using optical/lidar features in a central european landscape, *International Journal of Digital Earth* DOI:10.1080/17538947.2011.583992.

LeMay, V. & Temesgen, H. (2005). Camparison of nearest neighbour methods for estimating basal area and stems per hectare using aerial auxiliary variables, *Forest Science* 51(2): 109–119.

Liaw, A. & Wiener, M. (2002). Classification and regression by randomforest, *R News* 2: 18–22.

Lim, K., Treitz, P., Wulder, M., St-Onge, B. & Flood, M. (2003). Lidar remote sensing of forest structure, *Progress in Physical Geography* 27(1): 88–106.

MacLean, G. & Krabill, W. (1986). Gross merchantable timber volume estimation using an airborne lidar system, *Canadian Journal of Remote Sensing* 12: 7Ű18.

Maltamo, M., Bollandsås, O. M., Vauhkonen, J., Breidenbach, J., Gobakken, T. & E, N. (2010). Comparing different methods for prediction of mean crown height in norway spruce stands using airborne laser scanner data, *Forestry* 83(3): 257–268.

Maltamo, M. & Eerikäinen, K. (2001). The most similar neighbour reference in the yield prediction of pinus kesiya stands in zambia, *Silva Fennica* 35(4): 437–451.

Maltamo, M., Eerikäinen, K., Packalén, P. & Hyyppä, J. a. (2006). Estimation of stem volume using laser scanning-based canopy height metrics, *Forestry* 79(2): 217–229.

Maltamo, M., Hyyppä, J. & Malinen, J. (2006). A comparative study of the use of laser scanner data and field measurements in the prediction of crown height in boreal forests, *Scandinavian Journal of Forest Research* 21: 231–238.

Maltamo, M., Malinen, J., Packalén, P., Suvanto, A. & Kangas, J. (2006). Non-parametric estimation of stem volume using airborne laser scanning, aerial photography, and stand-register data, *Canadian Journal of Forest Research* 36: 426–436.

Maltamo, M., Næsset, E., Bollandsås, O., Gobakken, T. & Packalén, P. (2009). Non-parametric prediction of diameter distribution using airborne laser scanner data, *Scandinavian Journal of Forest Research* 24: 541–553.

McElhinny, C., Gibbons, P., Brack, C. & Bauhus, J. (2005). Forest and woodland stand structural complexity: Its definition and measurement, *Forest Ecology and Management* 218: 1–24.

McInerney, D. O., Suarez-Minguez, J., Valbuena, R. & Nieuwenhuis, M. (2010). Forest canopy height retrieval using lidar data, medium resolution satellite imagery and knn estimation in aberfoyle, scotland, *Forestry* 83(2): 195–206.

McRoberts, R. E. & Tomppo, E. O. (2007). Remote sensing support for national forest inventories, *Remote Sensing of Environment* 110: 412–419.

Mäkelä, H. & Pekkarinen, A. (2004). Estimation of forest stand volumes by landsat tm imagery and stand-level field-inventory data, *Forest Ecology and Managament* 196: 245–255.

Moeur, M. & Stage, A. R. (1995). Most similar neighbour: An improved sampling inference procedure for natural resource planning, *Forest Science* 41: 337Ű359.

Mohammadi, J. & Shataee, S. (2010). Possibility investigation of tree diversity mapping using landsat etm+ data in the hyrcanian forests of iran, *Remote Sensing of Environment* 104(7): 1504–1512.

Muukkonen, P. & Heiskanen, A. J. (2007). Biomass estimation over a large area based on standwise forest inventory data and aster and modis satellite data: A possibility to verify carbon inventories, *Remote Sensing of Environment* 107: 607–624.

Nelson, R., Krabill, W. & Tonelli, J. (1988). Estimating forest biomass and volume using airborne laser scanner data, *Remote Sensing of Environment* 24(2): 247–267.

Nothdurft, A., Soborowski, J. & Breidenbach, J. (2009). Spatial prediction of forest stand variables, *European Journal of Forest Research* 128(3): 241–251.

Næsset, E. (2002). Predicting forest stand characteristics with airborne scanning laser using a practical two-stage procedure and field data, *Remote Sensing of Environment* 80(1): 88–99.

Næsset, E. (2004). Practical large-scale forest stand inventory using a small airborne scanning laser, *Scandinavian Journal of Forest Research* 19: 164–179.

Oliver, C. & Larson, B. (1996). *Forest Stand Dynamics*, McGraw-Hill Inc., New York.

Packalén, P. & Maltamo, M. (2006). Predicting the plot volume by tree species using airborne laser scanning and aerial photographs, *Forest Science* 52(6): 611–622.

Packalén, P. & Maltamo, M. (2007). The k-msn method for the prediction of species-specific stand attributes using airborne laser scanning and aerial photographs, *Remote Sensing of Environment* 109: 328–341.

Packalén, P. & Maltamo, M. (2008). Estimation of species-specific diameter distributions using airborne laser scanning and aerial photographs, *Canadian Journal of Forest Research* 38: 1750–1760.

Pesonen, A., Maltamo, M., Packalén, P. & Eerikäinen, K. (2008). Airborne laser scanning-based prediction of coarse woody debris volumes in a conservation area, *Forest Ecology and Management* 255: 3288–3296.

Päivinen, R., Van Brusselen, J. & Schuck (2009). A the growing stock of european forests using remote sensing and forest inventory data, *Forestry* 82(5): 479–490.

Rahman, M., Csaplovics, E. & Koch, B. (2007). An efficient regression strategy for extracting forest biomass information from satellite sensor data, *International Journal of Remote Sensing* 26(7): 1511–1519.

Schlerf, M., Atzberger, C. & Hill, J. (2005). Remote sensing of forest biophysical variables using hymap imaging spectrometer data, *Remote Sensing of Environment* 95(2): 177–194.

Sexton, J. O., Bax, T., Siquiera, P., Swenson, J. J. & Hensley, S. (2009). comparison of lidar, radar, and field measurements of canopy height in pine and hardwood forests of southeastern north america, *Forest Ecology and Management* 257: 1136Ű1147.

Stage, A. R. & Crookston, N. L. (2007). Partitioning error components for accuracy-assessment of near- neighbor methods of imputation, *Forest Science* 53(1): 62?72.

Stümer, W. . D. (2004). *Kombination vor terrestischen Aufnahmen und Fernerkundungsdaten mit Hilfe der kNN-Methode zur Klassifizierung und Kartierung von Wäldern*, PhD thesis, Fakultät für Forst-, Geo- und Hydrowissenschaften der Technischen Universität Dresden.

Stoffels, J. (2009). *Einsatz einer lokal adaptiven Klassifikationsstrategie zur satellitengestützten Waldinventur in einem heterogenen Mittelgebirgsraum.*, PhD thesis, Faculty of Geography/Geesciences, University of Trier.

Stone, J. & Porter, J. (1998). What is forest stand structure and how to measure it?, *Northwest Science* 72(2): 25–26.

Straub, C., Dees, M., Weinacker, H. & Koch, B. (2009). Using airborne laser scanner data and cir orthophotos to estimate the stem volume of forest stands, *Photogrammetrie, Fernerkundung, GeoInformation* 3/2009: 277–287.

Straub, C. & Koch, B. (2011). Estimating single tree stem volume of pinus sylvestris using airborne laser scanner and multispectral line scanner data, *Remote Sensing* 3(5): 929–944.

Straub, C., Weinacker, H. & Koch, B. (2010). A comparison of different methods for forest resource estimation using information from airborne laser scanning and cir orthophotos, *European Journal of Forest Research* 129: 1069–1080.

Thessler, S., Sesnie, S., Bendana, Z., Ruokolainen, K., Tomppo, E. & Finegan, B. (2008). Using k-nn and discriminant analyses to classify rain forest types in a landsat tm image over northern costa rica, *Remote Sensing of Environment* 112: 2485– 2494.

Tommola, M., Tynkkynen, M., Lemmetty, J., Herstela, P. & Sikanen, L. (1999). Estimating the characteristics of a marked stand using k-nearest- neighbour regression, *Journal of Forest Engineering* pp. 75–81.

Tomppo, E. (1991). Satellite image-based national forest inventory of finland, *International Archives of Photogrammetry and Remote Sensing* 28 (7-1): 419Ŭ 424.

Tomppo, E. (1993). Multi-source national forest inventory of finland, *in* J. R. A. Nyyssoĺnen, S. Poso (ed.), *Proceedings of Ilvessalo symposium on national forest inventories*, p. 53 Ŭ 61.

Tomppo, E., Gagliano, C., De Natale, F., Katila, M. & McRoberts, R. E. (2009). Predicting categorical forest variables using an improved k-nearest neighbour estimator and landsat imagery, *Remote Sensing of Environment* 113(3): 500–517.

Tomppo, E. & Halme, M. (2004). Using coarse scale forest variables as ancillary information and weighting of variables in k-nn estimation: a genetic algorithm approach, *Remote Sensing of Environment* 92: 1–20.

Tomppo, E., Korhonen, K. T., Heikkinen, J. & Yli-Kojola, H. (2001). Multi-source inventory of the forests of the hebei forestry bureau, heilongjiang, china, *Silva Fennica* 35(3): 309Ŭ328.

Treuhaft, R. N., Asner, G. P. & Law, B. E. (2003). Structure-based forest biomass from fusion of radar and hyperspectral observations, *Geophysical Research Letters* 30(9): 1472.

Tyrrell, L. & Crow, T. (1994). Structural characteristics of old-growth hemlock-hardwood forests in relation to age, *Ecology* 75(2): 370–386.

Uuttera, J., Maltamo, M. & Hotanen, J. (1997). The structure of forest stands in virgin and managed peat-lands: a comparison between finnish and russian keralia, *Forest Ecology and Management* 96: 125–138.

Van Den Meersschaut, D. & Vandekerkhove, K. (1998). Development of a standscale forest biodiversity index based on the state forest inventory, *in* M. Hansen & T. Burk (eds), *Integrated Tools for Natural Resources Inventories in the 21st Century*, USDA, Boise, Idaho, USA, pp. 340–34.

Vauhkonen, J., Korpela, I., Maltamo, M. & Tokola, T. (2010). Imputation of single-tree attributes using airborne laser scanning-based height, intensity, and alpha shape metrics, *Remote Sensing of Environment* 114: 1263–1276.

Vohland, M., Stoffels, J., Hau, C. & Schüler, G. (2007). Remote sensing techniques for forest parameter assessment: Multispectral classification and linear spectral mixture analysis, *Silva Fennica* 41(3): 441–456.

Wallerman, J. & Holmgren, J. (2007). Estimating field-plot data of forest stands using airborne laser scanning and spot hrg data, *Remote Sensing of Environment* 110: 501–508.

Wehr, A. & Lohr, O. (1999). Airborne laser scanningŬan introduction and overview, *ISPRS Journal of Photogrammetry and Remote Sensing* 54: 68–82.

Wood, S. (2006). *Generalized additive models: an introduction with R*, Chapman & Hall/CRC, Boca Raton, Florida.

Yu, X., Hyyppä, J., Holopainen, M. & Vastaranta, M. . . . (2010). Comparison of area-based and individual tree-based methods for predicting plot-level forest attributes, *Remote Sensing* 2: 1481–1495.

Yu, X., Hyyppä, J., Vstarana, M., Holopainen, M. & Viitala, R. (2011). Predicting individual tree attributes from airborne laser point clouds based on the random forests technique, *ISPRS Journal of Photogrammetry and Remote Sensing* 66(1): 28–37.

3

Classification of Pre-Filtered Multichannel Remote Sensing Images

Vladimir Lukin[1], Nikolay Ponomarenko[1], Dmitriy Fevralev[1],
Benoit Vozel[2], Kacem Chehdi[2] and Andriy Kurekin[3]

[1]National Aerospace University
[2]University of Rennes 1
[3]Plymouth Marine Laboratory
[1]Ukraine
[2]France
[3]UK

1. Introduction

Multichannel remote sensing (RS) has gained popularity and has been successfully applied for solving numerous practical tasks as forestry, agriculture, hydrology, meteorology, ecology, urban area and pollution control, etc. (Chang, 2007). Using the term "multichannel", we mean a wide set of imaging approaches and RS systems (complexes) including multifrequency and dual/multi polarization radar (Oliver & Quegan, 2004), multi- and hyperspectral optical and infrared sensors. While for such radars the number of formed images is a few, the number of channels (components or sub-bands) in images can be tens, hundreds and even more than one thousand for optical/infrared imagers. TerraSAR-X is a good example of modern multichannel radar system; AVIRIS, HYDICE, HYPERION and others can serve as examples of modern hyperspectral imagers, both airborne and spaceborne (Landgrebe, 2002; Schowengerdt, 2007).

An idea behind increasing the number of channels is clear and simple: it is possible to expect that more useful information can be extracted from more data or this information is more reliable and accurate. However, the tendency to increasing the channels' (sub-band) number has also its "black" side. One has to register, to process, to transmit and to store more data. Even visualization of the obtained multichannel images for their displaying at tristimuli monitors becomes problematic (Zhang et al., 2008). Huge size of the obtained data leads to difficulties at any standard stage of multichannel image processing involving calibration, georeferencing, compression if used (Zabala et al., 2006). But, probably, the most essential problems arise in image pre-filtering and classification.

The complexity of these tasks deals with the following:

a. Noise characteristics in multichannel image components can be considerably different in the sense of noise type (additive, multiplicative, signal-dependent, mixed), statistics (probability density function (PDF), variance), spatial correlation (Kulemin et al., 2004; Barducci et al., 2005, Uss et al., 2011, Aiazzi et al., 2006);

b.   These characteristics can be a priori unknown or known only partly, signal-to-noise ratio can considerably vary from one to another component image (Kerekes & Baum, 2003) and even from one to another data cube of multichannel data obtained for different imaging missions;

c.   Although there are numerous books and papers devoted to image filter design and performance analysis (Plataniotis &Venetsanopoulos, 2000; Elad, 2010), they mainly deal with grayscale and color image processing; there are certain similarities between multichannel image filtering and color image denoising but the former case is sufficiently more complicated;

d.   Recently, several papers describing possible approaches to multichannel image filtering have appeared (De Backer et al., 2008; Amato et al., 2009; Benedetto et al., 2010; Renard et al., 2006; Chen & Qian, 2011; Demir et al., 2011, Pizurica & Philips, 2006; Renard et al., 2008); a positive feature of some of these papers is that they study efficiency of denoising together with classification accuracy; this seems to be a correct approach since classification (in wide sense) is the final goal of multichannel RS data exploitation and filtering is only a pre-requisite for better classification; there are two main drawbacks of these papers: noise is either simulated and additive white Gaussian noise (AWGN) is usually considered as a model, or aforementioned peculiarities of noise in real-life images are not taken into account;

e.   Though efficiency of filtering and classification are to be studied together, there is no well established correlation between quantitative criteria commonly used in filtering (and lossy compression) as mean square error (MSE), peak signal-to-noise ratio (PSNR) and some others and criteria of classification accuracy as probability of correct classification (PCC), misclassification matrix, anomaly detection probability and others (Christophe et al., 2005);

f.   One problem in studying classification accuracy is availability of numerous classifiers currently applied to multichannel images as neural network (NN) ones (Plaza et al., 2008), Support Vector Machines (SVM) and their modifications (Demir et al., 2011), different statistical and clustering tools (Jeon & Landgrebe, 1999), Spectral Angle Mapper (SAM) (Renard et al., 2008), etc.;

g.   It is quite difficult to establish what classifier is the best with application to multichannel RS data because classifier performance depends upon many factors as methodology of learning, parameters (as number of layers and neurons in them for NN), number of classes and features' separability, etc.; it seems that many researchers are simply exploiting one or two classifiers that are either available as ready computer tools or for which the users have certain experience;

h.   Dimensionality reduction, especially for hyperspectral data, is often used to simplify classification, to accelerate learning, to avoid dealing with spectral bands for which signal-to-noise ratios (SNRs) are quite low (Chen & Qian, 2011) due to atmospheric effects; to exploit only data from those sub-bands that are the most informative for solving a given particular task (Popov et al.., 2011); however, it is not clear how to perform dimensionality reduction in an optimal manner and how filtering influences dimensionality reduction;

i.   Test multichannel images for which it could be possible to analyze efficiency of filtering and accuracy of classification are absent; because of this, people either add noise of quite high level to real-life data (that seem practically noise free) artificially or characterize efficiency of denoising by the "final result", i.e. by increasing the PCC (Chen & Qian, 2011).

It follows from the aforesaid that it is impossible to take into account all factors mentioned above. Thus, it seems reasonable to concentrate on considering several particular aspects. Therefore, within this Chapter we concentrate on analyzing multichannel data information component and noise characteristics first. To our opinion, this is needed for better understanding of what are peculiarities of requirements to filtering and what approaches to denoising can be applied. All these questions are thoroughly discussed in Section 2 with taking into account recent advances in theory and practice of image filtering. Besides, we briefly consider some aspects of classifier training in Section 3. Section 4 deals with analysis of classification results for three-channel data created on basis of Landsat images ·with artificially added noise. Throughout the Chapter, we present examples from real-life RS images of different origin to provide generality of analysis and conclusions.

One can expect that more efficient filtering leads to better classification. This expectation is, in general, correct. However, considering image filtering, one should always keep in mind that alongside with noise removal (which is a positive effect) any filter produces distortions and artefacts (negative effects) that influence RS data classification as well. Because of this, filtering, to be reasonable for applying, has to provide more positive effects than negative ones from the viewpoint of solving a final task, RS data classification in the considered case.

2. Approaches to multichannel image filtering

2.1 Information content and noise characteristics

Speaking very simply, benefits of multichannel remote sensing compared to single-channel mode are due to the following reasons. First, availability of multichannel (especially hyperspectral) data allows solving many particular tasks since while for one particular task one subset of sub-band data is "optimal", another subset is "optimal" for solving another task. Thus, multichannel remote sensing is multi-purpose allowing different users to be satisfied with employing data collected one time for a given territory. Second, useful information is often extracted by exploiting certain similarity of information content in component images and practical independence of noise in these components. Thus, efficient SNR increases due to forming and processing more sub-band images.

Really, correlation of information content in multichannel RS data is usually high. Let us give one example. Consider hyperspectral data provided by AVIRIS airborne system (available at http://aviris.jpl.nasa.gov/aviris) that can be represented as $I(i, j, \lambda)$ where $i = 1, ..., I_{im}, j = 1, ..., J_{im}$ denote image size and λ is wavelength, $\lambda_k, k = 1, ... K$ defines wavelength for a k-th subband (the total number of sub-bands for AVIRIS images is 224). Let us analyze cross-correlation factors determined for neighbouring k-th and $k+1$-th sub-band images as

$$R^{k\,k+1} = (\sum_{i=1}^{I_{im}} \sum_{j=1}^{J_{im}} (I(i, j, \lambda_k) - I_{mean}(\lambda_k))(I(i, j, \lambda_{k+1}) - I_{mean}(\lambda_{k+1}))) / (I_{im} J_{im} \sigma_k \sigma_{k+1}) \quad (1)$$

$$I_{mean}(\lambda_k) = \sum_{i=1}^{I_{im}} \sum_{j=1}^{J_{im}} I(i, j, \lambda_k) / (I_{im} J_{im})$$

$$\sigma_k^2 = \sum_{i=1}^{I_{im}} \sum_{j=1}^{J_{im}} (I(i,j,\lambda_k) - I_{mean}(\lambda_k))^2 / (I_{im}J_{im} - 1)$$

The obtained plot for AVIRIS data is presented in Fig. 1. It is seen that for most neighbour sub-bands the values of R^{kk+1} are close to unity confirming high correlation (very similar content) of these images. There are such k for which R^{kk+1} considerably differs from unity. In particular, this happens for several first sub-bands, several last sub-bands, sub-bands with k about 110 and 160. The main reason for this is the presence of noise.

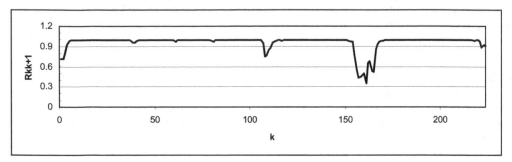

Fig. 1. The plot $R^{kk+1}, k = 1,...,223$ for AVIRIS image Moffett Field

To prove this, let us present data from the papers (Ponomarenko et al., 2006) and (Lukin et al., 2010b). Based on blind estimates of additive noise standard deviations in sub-band images $\hat{\sigma}_{adk}^2$, robust modified estimates PSNR$_{mod}$ have been obtained for all channels (modifications have been introduced for reducing the influence of hot pixel values):

$$PSNR_{mod}(k) = 10\log_{10}((I_{99\%}(k) - I_{1\%}(k))^2 / \hat{\sigma}_{adk}^2 \tag{2}$$

where $I_{q\%}(k)$ defines q-th percent quintile of image values in k-th sub-band image.

The plot is presented in Fig. 2. Comparing the plots in Figures 1 and 2, it can be concluded that rather small R^{kk+1} are observed for such subintervals of k for which $PSNR_{mod}(k)$ are also quite small. Thus, there is strict relation between these parameters.

There is also relation between $PSNR_{mod}(k)$ and SNR for sub-band images analyzed in Ref. (Curran & Dungan, 1989). In this sense, one important peculiarity of multichannel (especially, hyperspectral) data is to be stressed. Dynamic range of the data in sub-band images characterized by $I_{max}(k) - I_{min}(k)$ (maximal and minimal values for a given k-th sub-band) varies a lot. Note that to avoid problems with hot pixels and outliers in data, it is also possible to characterize dynamic range by $I_{99\%}(k) - I_{1\%}(k)$ exploited in (2).

The plot of $D_{rob}(k) = I_{99\%}(k) - I_{1\%}(k)$ is presented in Fig. 3. It follows from its analysis that a general tendency is decreasing of $D_{rob}(k)$ when k (and wavelength) increases with having sharp jumps down for sub-bands where atmospheric absorption and other physical effects take place. Though both $PSNR_{mod}(k)$ and SNR can characterize noise influence (intensity) in images, we prefer to analyze $PSNR_{mod}(k)$ and $PSNR$ below as parameters more commonly used in practice of filter efficiency analysis. Strictly saying, $PSNR_{mod}(k)$ differs

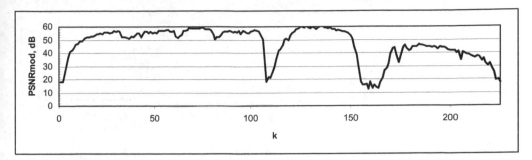

Fig. 2. $PSNR_{mod}(k)$ for the same image as in Fig. 1

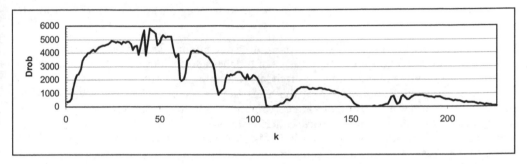

Fig. 3. Robustly estimated dynamic range $D_{rob}(k)$

from traditional PSNR, but for images without outliers this difference is not large and the tendencies observed for *PSNR* take place for $PSNR_{mod}(k)$ as well.

Noise characteristics in multichannel image channels can be rather different as well. The situation when noise type is different happens very seldom (this is possible if, e.g., optical and synthetic aperture radar (SAR) data are fused (Gungor & Shan, 2006) where additive noise model is typical for optical data and multiplicative noise is natural for radar ones). The same type of noise present in all component images is the case met much more often. However, noise type can be not simple and noise characteristics (e.g., variance) can change in rather wide limits. Let us give one example. The estimated standard deviation (STD) of additive noise for all sub-band images is presented in Fig. 4. As it is seen, the estimates vary a lot. Even though these are estimates with a limited accuracy, the observed variations clearly demonstrate that noise statistics is not constant.

A more thorough analysis (Uss et al., 2011) shows that noise is not purely additive but signal dependent even for data provided by such old hyperspectral sensors as AVIRIS. Sufficient variations of signal dependent noise parameters from one band to another are observed. Recent studies (Barducci et al., 2005, Alparone et al., 2006) demonstrate a clear tendency for signal-dependent noise component to become prevailing (over additive one) for new generation hyperspectral sensors. This means that special attention should be paid to this tendency in filter design and efficiency analysis with application to multichannel data denoising and classification. Although the methods of multichannel image denoising designed on basis of the additive noise model with identical variance in all component

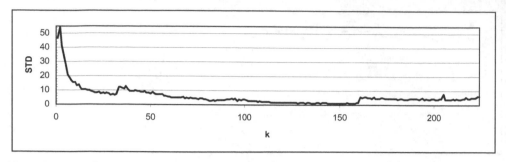

Fig. 4. Estimated STD of noise for components of the same AVIRIS image

images can provide a certain degree of noise removal, they are surely not optimal for the considered task.

Consider one more example. Figure 5 presents two components of dual-polarisation (HH and VV) 512x512 pixel fragment SAR image of Indonesia formed by TerraSAR-X spaceborne system (http://www.infoterra.de/tsx/freedata/start.php). Amplitude images are formed from complex-valued data offered at this site. As it is seen, the HH and VV images are similar to each other although both are corrupted by fully developed speckle and there are some differences in intensity of backscattering for specific small sized objects placed on water surface (left part of images, dark pixels). The value of cross-correlation factor (1) is equal to 0.63, i.e. it is quite small. Both images have been separately denoised by the DCT-based filter adapted to multiplicative nature of noise (with the same characteristics for both images) and spatial correlation of speckle (Ponomarenko et al., 2008a). The filtered images are represented in Fig. 6 where it is seen that speckle has been effectively suppressed. Filtering has considerably increased inter-channel correlation, it is equal to 0.85 for denoised images. This indirectly confirms that low values of inter-channel correlation factor in original RS data can be due to noise.

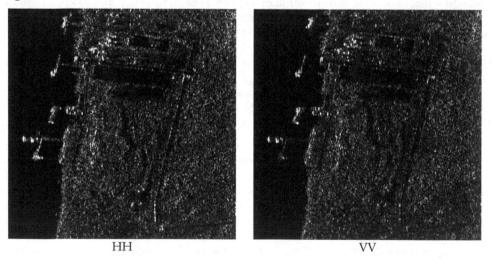

HH   VV

Fig. 5. The 512x512 pixel fragment SAR images of Indonesia for two polarizations

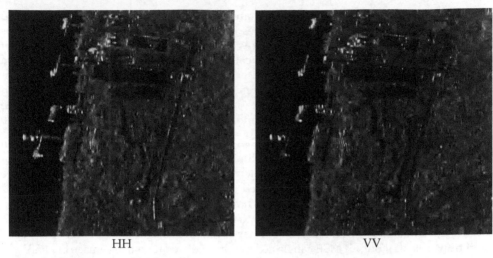

HH   VV

Fig. 6. The SAR images after denoising

The given example for dual polarization SAR data is also typical in the sense that noise in component images can be not additive (speckle is pure multiplicative) and not Gaussian (it has Rayleigh distribution for the considered amplitude single look SAR images). For the presented example of HH and VV polarization images statistical and spatial correlation characteristics of speckle are practically identical in both component images, but it is not always the case for multichannel radar images.

The presented results clearly demonstrate that noise in multichannel RS images can be signal-dependent where its variance (and sometimes even PDF) depends upon information signal (image). Noise statistics can also vary from one sub-band image to another. These peculiarities have to be taken into account in multichannel image simulation, filter and classifier design and performance analysis.

2.2 Component-wise and vector filtering

If one deals with 3D data as multichannel RS images, an idea comes immediately that filtering can be carried out either component-wise or in a vector (3D) manner. This was understood more than 20 years ago when researchers and engineers ran into necessity to process colour RGB images (Astola et al.,1990). Whilst for colour images there are actually only these two ways, for multichannel images there is also a compromise variant of processing not entire 3D volume of data but also certain groups (sets) of channels (sub-bands) (Uss et al., 2011). As analogue of this situation, we can refer to filtering of video where a set of subsequent frames can be used for denoising (Dabov et al., 2007). There is also possibility to apply denoising only to some but not all component images. In this sense, it is worth mentioning the paper (Philips et al., 2009). It is demonstrated there that pre-filtering of some sub-band images can make them useful for improving hyperspectral data classification carried out using reduced sets of the most informative channels. However, the proposed solution to apply the median filter with scanning windows of different size component-wise is, to our opinion, not the best choice.

Thus, there are quite many opportunities and each way has its own advantages and drawbacks. Keeping in mind the peculiarities of image and noise discussed above, let us start from the simplest case of component-wise filtering. It is clear that more efficient filtering leads, in general, to better classification (although strict relationships between conventional quantitative criteria characterizing filtering efficiency and classifier performance are not established yet). Therefore, let us revisit recent achievements and advances in theory and practice of grayscale image filtering and analyze in what degree they can be useful for hyperspectral image denoising.

Recall that the case of additive white Gaussian noise (AWGN) present in images has been studied most often. Recently, the theoretical limits of denoising efficiency in terms of output mean square error (MSE) within non-local filtering approach have been obtained (Chatterjee & Milanfar, 2010). The authors have presented results for a wide variety of test images and noise variance values. Moreover, the authors have provided software that allows calculating potential (minimal reachable) output MSE for a given noise-free grayscale image for a given standard deviation of AWGN. Later, in the paper (Chatterjee & Milanfar, 2011), it has been shown how potential output MSE can be accurately predicted for a noisy image at hand.

This allows drawing important conclusions as follows. First, potential reduction of output MSE compared to variance of AWGN in original image depends upon image complexity and noise intensity. Reduction is large if an image is quite simple and noise variance is large, i.e. if input SNR (and PSNR) of an image to be filtered is low. For textural images and high input SNR, potential output MSE can be by only 1.2...1.5 times smaller than AWGN variance (see also data in the papers (Lukin at al., 2011, Ponomarenko et al., 2011, Fevralev et al., 2011)). This means that filtering becomes practically inefficient in the sense that positive effect of noise removal is almost "compensated" by negative effect of distortion introducing inherent for any denoising method in less or larger degree. With application to hyperspectral data filtering, this leads to the aforementioned idea that not all component images are to be filtered. The preliminary conclusion then is that sub-band images with rather high SNR are to be kept untouched whilst other ones can be denoised. A question is then what can be (automatic) rules for deciding what sub-band images to denoise and what to remain unfiltered? Unfortunately, such rules and automatic procedures are not proposed and tested yet. As preliminary considerations, we can state only that if input PSNR is larger than 35 dB, then it is hard to provide PSNR improvement due to filtering by more than 2...3 dB. Moreover, for input PSNR>35 dB, AWGN in original images is almost not seen (it can be observed only in homogeneous image regions with rather small mean intensity). Because of this, denoised and original component images might seem almost identical (Fevralev et al., 2011). Then it comes a question is it worth carrying out denoising for such component images with rather large input PSNR in the sense of filtering positive impact on classification accuracy. We will turn back to this question later in Section 4.

The second important conclusion that comes from the analysis in (Chatterjee & Milanfar, 2010) is that the best performance for grayscale image filtering is currently provided by the methods that belong to the non-local denoising group (Elad, 2010; Foi et al., 2007; Kervrann & Boulanger, 2008). The best orthogonal transform based methods are comparable to non-local ones in efficiency, especially if processed images are not too simple (Lukin et al., 2011a). Let us see how efficient these methods can be with application to component-wise processing of multichannel RS data.

Although noise is mostly signal-dependent in component images of hyperspectral data, there are certain sub-bands where dynamic range is quite small and additive noise component is dominant or comparable to signal-dependent one (Uss et al., 2011; Lukin et al., 2011b). One such image (sub-band 221 of the AVIRIS data set Cuprite) is presented in Fig. 7,a. Noise is clearly seen in this image and the estimated variance of additive noise component is about 30. The output image for the BM3D filter (Foi et al., 2007) which is currently the best among non-local denoisers is given in Fig. 7,b. Noise is suppressed and all details and edges are preserved well.

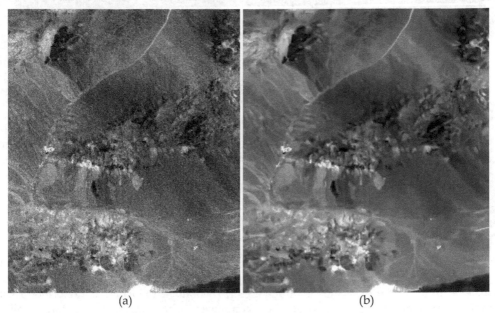

(a)   (b)

Fig. 7. Original 221 sub-band AVIRIS image Cuprite (a) and the output of BM3D filter (b)

However, applying the non-local filters becomes problematic if noise does not fit the (dominant) AWGN model considered above. There are several problems and few known ways out. The first problem is that the non-local denoising methods are mostly designed for removal of AWGN. Recall that these methods are based on searching for similar patches in a given image. The search becomes much more complicated if noise is not additive and, especially, if noise is spatially correlated. One way out is to apply a properly selected homomorphic variance-stabilizing transform to convert a signal dependent noise to pure additive and then to use non-local filtering (Mäkitalo et al., 2010). This is possible for certain types of signal-dependent noise (Deledalle et al., 2011, see also www.cs.tut.fi/~foi/optvst). Thus, the considered processing procedure becomes applicable under condition that the noise in an image is of known type, its characteristics are known or properly (accurately) pre-estimated and there exists the corresponding pair of homomorphic transforms. Examples of signal dependent noise types for which such transforms exist are pure multiplicative noise (direct transform is of logarithmic type), Poisson noise (Anscombe transform), Poisson and pure additive noise (generalized Poisson transform) and other ones.

Let us demonstrate applicability of the three-stage filtering procedure (direct homomorphic transform – non-local denoising – inverse homomorphic transform) for noise removal in SAR images corrupted by pure multiplicative noise (speckle). The output of this procedure exploited for processing the single-look SAR image in Fig. 5 (HH) is represented in Fig. 8,a. Details and edges are preserved well and speckle is sufficiently suppressed.

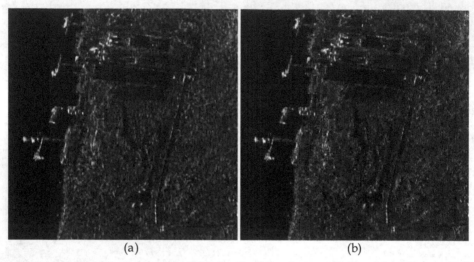

(a)   (b)

Fig. 8. The HH SAR image after denoising by the three-stage procedure (a) and vector DCT-based filtering (b)

The second problem is that similar patch search becomes problematic for spatially correlated noise. For correlated noise, similarity of patches can be due to similarity of noise realizations but not due to similarity of information content. Then, noise reduction ability of non-local denoising methods decreases and artefacts can appear. The problem of searching similar blocks (8x8 pixel patches) has been considered (Ponomarenko et al., 2010). But the proposed method has been applied to blind estimation of noise spatial spectrum in DCT domain, not to image filtering within non-local framework. The obtained estimates of the DCT spatial spectrum have been then used to improve performance of the DCT based filter (Ponomarenko2008). Note that adaptation to spatial spectrum of noise in image filtering leads to sufficient improvement of output image quality according to both conventional criteria and visual quality metrics (Lukin et al, 2008).

Finally, the third problem deals with accurate estimation of signal-dependent noise statistical characteristics (Zabrodina et al., 2011). Even assuming a proper variance stabilizing transform exists as, e.g., generalized Anscombe transform (Murtag et al, 1995) for mixed Poisson-like and additive noise, parameters of transform are to be adjusted to mixed noise statistics. Then, if statistical characteristics of mixed noise are estimated not accurately, variance stabilization is not perfect and this leads to reduction of filtering efficiency. Note that blind estimation of mixed noise parameters is not able nowadays to provide quite accurate estimation of parameters for all images and all possible sets of mixed noise parameters (Zabrodina et al., 2011). Besides, non-local filtering methods are usually not fast enough since search for similar patches requires intensive computations.

As an alternative solution to three-stage procedures that employ non-local filtering, it is possible to advice using locally adaptive DCT-based filtering (Ponomarenko et al., 2011). Under condition of a priori known or accurately pre-estimated dependence of signal dependent noise variance on local mean $\sigma_{sd}^2 = f(I^{tr})$, it is easy to adapt local thresholds for hard thresholding of DCT coefficients in each nm-th block as

$$T(n,m) = \beta \sqrt{f(\widehat{\bar{I}}(n,m))} \tag{3}$$

where $\widehat{\bar{I}}(n,m)$ is the estimate of the local mean for this block, β is the parameter (for hard thresholding, $\beta = 2.6$ is recommended). If noise is spatially correlated and its normalized spatial spectrum $W_{norm}(k,l)$ is known in advance or accurately pre-estimated, the threshold becomes also frequency-dependent

$$T(n,m,k,l) = \beta \sqrt{W_{norm}(k,l) f(\widehat{\bar{I}}(n,m))} \tag{4}$$

where k and l are frequency indices in DCT domain.

One more option is to apply the modified sigma filter (Lukin et al., 2011b) where the neighbourhood for a current ij-th pixel is formed as

$$I_{min}(i,j) = I(i,j) - \alpha_{sig}\sqrt{f(I(i,j))}, \, I_{max}(i,j) = I(i,j) + \alpha_{sig}\sqrt{f(I(i,j))}, \tag{5}$$

where α_{sig} is the parameter commonly set equal to 2 (Lee, 1980) and averaging of all image values for ij-th scanning window position that belong to the interval defined by (5) is carried out. This algorithm is very simple but not as efficient as the DCT-based filtering in the same conditions (Tsymbal et al., 2005). Moreover, the sigma filter can be in no way adapted to spatially correlated noise.

Finally, if there is no information on $\sigma_{sd}^2 = f(I^{tr})$ and $W_{norm}(k,l)$, it is possible to use an adaptive DCT-based filter version designed for removing non-stationary noise (Lukin et al., 2010a). However, for efficient filtering, it is worth exploiting all information on noise characteristics that is either available or can be retrieved from a given image.

Let us come now to considering possible approaches to vector filtering of multichannel RS data. Again, let us start from theory and recent achievements. First of all, it has been recently shown theoretically that potential output MSE for vector (3D) processing is considerably better (smaller) than for component-wise filtering of color RGB images (Uss et al., 2011b), by 1.6...2.2 times. This is due to exploiting inherent inter-channel correlation of signal components. Then, if a larger number of channel data are processed together and inter-channel correlation factor is larger than for RGB color images (where it is about 0.8), one can expect even better efficiency of 3D filtering.

Similar effects but concerning practical output MSEs have been demonstrated for 3D DCT based filter (Ponomarenko et al., 2008b) and vector modified sigma filter (Kurekin et al., 1999; Lukin et al., 2006; Zelensky et al., 2002) applied to color and multichannel RS images. It is shown in these papers that vector processing provides sufficient benefit in filtering efficiency (up to 2 dB) for the cases of three-channel image processing with similar noise intensities in component images. This, in turn, improves classification of multichannel RS data (Lukin et al., 2006, Zelensky et al., 2002).

However, there are specific effects that might happen if 3D filtering is applied without careful taking into account noise characteristics in component images (and the corresponding pre-processing). For the vector sigma filter, the 3D neighborhood can be formed according to (5) for any a priori known dependences $f(.)$ that can be individual for each component image. This is one advantage of this filter that, in fact, requires no pre-processing operations as, e.g., homomorphic transformations. Another advantage is that if noise is of different intensity in component images processed together, then the vector sigma filter considerably improves the quality of the component image(s) with the smallest SNR. A drawback is that filtering for other components is not so efficient. The aforementioned property can be useful for hyperspectral data for which it seems possible to enhance component images with low SNR by proper selection of other component images (with high SNR) to be processed jointly (in the vector manner). However, this idea needs solid verification in future.

For the 3D DCT-based filtering, two practical situations have been considered. The first one is AWGN with equal variances in all components (Fevralev et al., 2011). Channel decorrelation and processing in fully overlapping 8x8 blocks is applied. This approach provides 1...2 dB improvement compared to component-wise DCT-based processing of color images according to output PSNR and the visual quality metric PSNR-HVS-M (Ponomarenko et al., 2007). The second situation is different types of noise and/or different variances of noise in component images to be processed together. Then noise type has to be converted to additive by the corresponding variance stabilizing transforms and images are to be normalized (stretched) to have equal variances. After this, the 3D DCT based filter is to be applied. Otherwise, e.g., if noise variances are not the same, oversmoothing can be observed for component images with smaller variance values whilst undersmoothing can take place for components with larger variances. To illustrate performance of this method, we have applied it to dual-polarization SAR image composed of images presented in Fig. 5. Identical logarithmic transforms have been used first separately for each component to get two images corrupted by pure additive noise with equal variance values. Then, the 3D DCT based filtering with setting the frequency dependent thresholds as $T(k,l) = \beta \sigma_{adc} \sqrt{W_{norm}(k,l)}$ has been used where σ_{adc} denotes additive noise standard deviation after direct homomorpic transform. Finally, identical inverse homomorphic transforms have been performed for each component image. The obtained filtered HH component image is presented in Fig. 8,b. Speckle is suppressed even better than in the image in Fig 8,a and edge/detail preservation is good as well.

Note that vector filtering of multichannel images can be useful not only for more efficient denoising, but also for decreasing residual errors of image co-registration (Kurekin1997). Its application results in less misclassifications in the neighborhoods of sharp edges.

As it is seen, the DCT-based filtering methods use the parameter β that, in general, can be varied. Analysis of the influence of this parameter on filtering efficiency for the three-channel LandSat image visualized in RGB in Fig. 9 has been carried out in (Fevralev et al., 2010). Similar analysis, but for standard grayscale images, has been performed in (Ponomarenko et al., 2011). It has been established that an optimal value of β that provides maximal efficiency of denoising according to a given quantitative criteria depends upon a filtered image, noise intensity (variance for AWGN case), thresholding type, and a metric used. In particular, for hard thresholding which is the most popular and rather efficient, optimal β is usually slightly larger than 2.6 if an image is quite simple, noise is intensive and output PSNR or MSE are used as criteria (β_{opt}^{PSNR}). For complex images and small variance of noise (input PSNR>32..34 dB), β_{opt}^{PSNR} is usually slightly smaller than 2.6. Interestingly, if the visual quality metric PSNR-HVS-M (Ponomarenko et al., 2007) is employed as criterion of filtering efficiency, the corresponding optimal value is $\beta_{opt}^{PSNR-HVS-M} \approx 0.85\beta_{opt}^{PSNR}$ for all considered images and noise intensities. This means that if one wishes to provide better visual quality of filtered image, edge/detail/texture preservation is to be paid main attention (better preservation is provided if β is smaller).

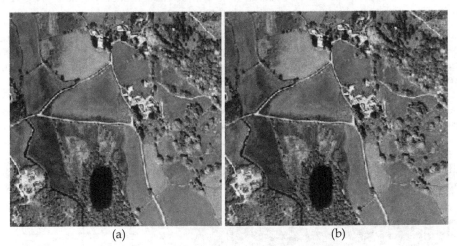

(a)   (b)

Fig. 9. Noise free (a) and noisy (b) test images, additive noise variance is equal to 100

3. Classifiers and their training

In this Section, we would like to avoid a thorough discussion on possible classification approaches with application to multichannel RS images. An interested reader is addressed to (Berge & Solberg, 2004), (Melgani & Bruzzone, 2004), (Ainsworth et al., 2007), etc. General observations of modern tendencies for hyperspectral images are the following. Although there are quite many different classifiers (see Introduction), neural network, support vector machine and SAM are, probably, the most popular ones. One reason for using NN and SVM classifiers is their ability to better cope with non-gaussianity of features. Dimensionality reduction (there are numerous methods) is usually carried out without loss in classification accuracy but with making the classification task simpler.

Classifier performance depends upon many factors as number of classes, their separability in feature space, classifier type and parameters, a methodology of training used and a training sample size, etc. If training is done in supervised manner (which is more popular for classification application), training data set should contain, at least, hundreds of feature vectors and classification is then carried out for other pixels (in fact, voxels or feature vectors obtained for them). Validation is usually performed for thousands of voxels. Pixel-by-pixel classification is usually performed, being quite complex even in this case, although some advanced techniques exploit also texture features (Rellier et al., 2004). There is also an opportunity to post-process preliminary classification data in order to partly remove misclassifications (Yli-Harja & Shmulevich, 1999).

The situation in classification of multichannel radar imagery is another due to considerably smaller number of channels (Ferro-Famil & Pottier, 2001, Alberga et al., 2008). There is no problem with dimensionality reduction. Instead, the problem is with establishing and exploiting sets of the most informative and noise-immune features derived from the obtained images. One reason is that there are many different representations of polarimetric information where features can be not independent, being retrieved from the same original data. Another reason is intensive speckle inherent for radar imagery where SARs able to provide appropriate resolution are mostly used nowadays.

To sufficiently narrow an area of our study, we have restricted ourselves by considering the three-channel Landsat image (Fig. 9a) composed of visible band images that relate to central wavelengths 0.66 µm, 0.56 µm, and 0.49 µm associated with R, G, and B components of the obtained "color" image, respectively. Only the AWGN case has been analyzed where noise with predetermined variance was artificially added to each component independently. Radial basis function (RBF) NN and SVM classifiers have been applied. According to the recommendations given above, training has been done for several fragments for each class shown by the corresponding colors in Fig. 10b. The numbers of training samples was 1617, 1369, 375, 191 and 722 for the classes "Soil", "Grass", "Water", "Urban" (Roads and Buildings), and "Bushes", respectively. Classification has been applied to all image pixels although validation has been performed only for pixels that belong to areas marked by five colors in Fig. 10a.

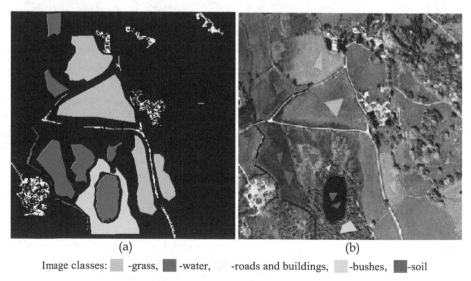

(a)   (b)

Image classes: ▨ -grass, ▨ -water,   -roads and buildings, ▨ -bushes, ▨ -soil

Fig. 10. Ground truth map (a) and fragments used for classifier training (b)

Pixel-by-pixel classification has been used without exploiting any textural features since these features can be influenced by noise and filtering. The training dataset has been formed from noise-free samples of the original test image represented in Fig. 9,a, to alleviate these impairments degrading the training results and to make simpler the analysis of image classification accuracy in the presence of noise and distortions introduced by denoising. Thus, in fact, for every image pixel the feature vector has been formed as $\mathbf{x}_q = \left(x_q^R, x_q^G, x_q^B \right)$, i.e. composed of brightness values of Landsat image components associated with R, G, and B.

Details concerning training the considered classifiers can be found in (Fevralev et al., 2010). Here we would like to mention only the following. We have used the RBF NN with one hidden layer of nonlinear elements with a Gaussian activation function (Bose & Liang, 1996) and an output layer with linear elements. The element number in the output layer equals to the number of classes (five) where every element is associated with the particular class of the sensed terrain. The classifier presumes making a hard decision that is performed by selecting the element of the output layer having the maximum output value. The RBF NN unknown parameters have been obtained by the cascade-correlation algorithm that starts with one hidden unit and iteratively adds new hidden units to reduce (minimize) the total residual error. The error function has exploited weights to provide equal contributions from every image class for different numbers of class learning samples.

The considered SVM classifier employs nonlinear kernel functions in order to transform a feature vector into a new feature vector in a higher dimension space where linear classification is performed (Schölkopf et al., 1999). The SVM training has been based on quadratic programming, which guarantees reaching a global minimum of the classifier error function (Cristianini & Shawe-Taylor, 2000). For the considered classification task, we have applied a Radial Basis kernel function of the same form as the activation function of the RBF NN hidden layer units. To solve multi-class problem using the SVM classifier we have applied one-against-one classification strategy. It divides the multi-class problem into $S(S-1)/2$ separate binary classification tasks for all possible pair combinations of S classes. A majority voting rule has been then applied at the final stage to find the resulting class.

The overall probability of correct classification reached for noise-free image is 0.906 for the RBF NN and 0.915 for the SVM classifiers, respectively. The reasons of the observed misclassifications are that the considered classes are not separable as we exploited only three simple features (intensities in channel images). The largest misclassification probabilities have been observed for the classes "Soil" and "Urban", "Soil" and "Bushes". This is not surprising since these classes are quite heterogeneous and have similar "colors" in the composed three-channel image (see Fig. 9,a).

4. Filtering and classification results and examples

Concerning Landsat data classification, let us start with considering overall probabilities of correct classification P_{cc}. The obtained results are presented in Table 1 for three values of AWGN variance, namely, 100, 49, and 16 (note that only two values, 100 and 49, have been analyzed in the earlier paper (Fevralev et al., 2010). The case of noise variance equal to 16 is added to study the situation when input PSNR=39 dB, i.e. noise intensity is such that noise

Image	σ^2	β	P_{cc} for SVM	P_{cc} for RBF NN
Noisy	16	-	0.890	0.887
Filtered (component-wise, HT)	16	2.5	0.909	0.905
Filtered (3D, HT)	16	2.5	0.919	0.906
Noisy	49	-	0.813	0.838
Filtered (component-wise, HT)	49	2.5	0.889	0.903
Filtered (component-wise, HT)	49	2.1	0.880	0.898
Filtered (component-wise, CT)	49	3.9	0.888	0.903
Filtered (component-wise, CT)	49	3.3	0.879	0.896
Filtered (3D, HT)	49	2.6	0.917	0.911
Noisy	100	-	0.729	0.766
Filtered (component-wise, HT)	100	2.5	0.881	0.902
Filtered (component-wise, HT)	100	2.1	0.867	0.892
Filtered (component-wise, CT)	100	3.9	0.879	0.902
Filtered (component-wise, CT)	100	3.3	0.865	0.890
Filtered (3D, HT)	100	2.6	0.918	0.914

Table 1. Classification results for original and filtered images

is practically not seen in original image (Fevralev et al., 2011). Alongside with hard thresholding (HT), we have analyzed a combined thresholding (CT)

$$D_{ct}(n,m,k,l) = \begin{cases} D(n,m,k,l), if \, |D(n,m,k,l)| \geq \beta\sigma(n,m,k,l) \\ D^3(n,m,k,l) \, / \, \beta^2\sigma^2(n,m,k,l) \, otherwise \end{cases} \quad (6)$$

where $\sigma^2(n,m,k,l) = f(\overline{I}(n,m)W_{norm}(k,l))$. Note that for CT $\beta_{opt}^{PSNR} \approx 3.9$ and the aforementioned property $\beta_{opt}^{PSNR-HVS-M} \approx 0.85\beta_{opt}^{PSNR}$ is also valid.

As it follows from analysis of data in Table 1, any considered method of pre-filtering noisy images has positive effect on classification irrespectively to a classifier used. As it could be expected, the largest positive effect associated with considerable increase of P_{cc} is observed if noise is intensive (see data for $\sigma^2=100$ compared to „Noisy"). If noise variance is small ($\sigma^2=16$), there is still improvement of image quality after filtering. Output PSNR becomes 42.4 dB after component-wise denoising and 43.0 after 3D DCT-based filtering. This improvement in terms of PSNR leads to increase of P_{cc} although it is not large. Probability of correct classification has sufficiently increased for classes 1 (Soil), 2 (Grass), and 5 (Bushes).

Note that for filtered image P_{cc} is practically the same as for classification of noise-free data. This shows that if PSNR for classified image is over 42...43 dB, the (residual) noise practically does not effect classification.

Both considered algorithms of thresholding produce approximately the same results for the same noise variance, classifier and component-wise filtering (compare, e.g., the cases

$\beta = 2.5$ for HT and $\beta = 3.9$ for CT, σ^2=100 and 49). Because of this, we have analyzed only hard thresholding for σ^2=16.

The use of smaller $\beta = 2.1$ for HT and $\beta = 3.3$ for CT (that correspond to $\beta_{opt}^{PSNR-HVS-M}$) results in slight reduction of P_{cc} compared to the case of setting β_{opt}^{PSNR} . To our opinion, this can be explained by better noise suppression efficiency provided for the DCT-based filtering with larger β which is expedient for, at least, two classes met in the studied Landsat image (namely, for „homogeneous" classes „Water" and „Grass" that occupy about half of pixels in validation set, see Fig. 10b). Data analysis also allows concluding that more efficient filtering provided by the 3D filtering compared to component-wise processing leads to sufficient increase in P_{cc} especially for intensive noise case and SVM classifier. This shows that if filtering is more efficient in terms of conventional metrics, then, most probably, it is more expedient in terms of classification. All these conclusions are consistent for both classifiers. Although the results are slightly better for the RBF NN if noise is intensive, P_{cc} values are almost the same for non-intensive noise.

We have also analyzed the influence of filtering efficiency on classification accuracy for particular classes. Only hard thresholding has been considered (the results for combined thresholding are given in (Fevralev et al., 2010) and they are quite close to the data for hard thresholding). Three filtering approaches have been used: component-wise denoising with $\beta_{opt}^{PSNR-HVS-M} = 2.1$ (denoted as Filtered 2.1), component-wise filtering with β_{opt}^{PSNR} (denoted as Filtered 2.5), and 3D (vector) processing (Filtered 3D).

For the first class "Soil", a clear tendency is observed: more efficient the filtering, larger the probability of correct classification P_{corr1}. The same holds for "homogeneous" classes "Grass" (analyze P_{corr2}) and "Water" (see data for P_{corr3}), the attained probabilities for these classes are high and approach unity for filtered images. The dependences for the class "Bushes" (see P_{corr5}) are similar to the dependences for the class "Soil". P_{corr5} increases if more efficient filtering is applied but not essentially. Quite many misclassifications remain due to "heterogeneity" of the classes "Soil" and "Bushes" (see discussion above).

Finally, specific results are observed for the class "Urban" (see data for P_{corr4}). The pixels that belong to this class are not classified well in noisy images, especially by the SVM classifier. Filtering, especially 3D processing that possesses the best edge/detail preservation, slightly improves the values of P_{corr4}. There is practically no difference in data for the cases Filtered 2.1 and Filtered 2.5.

Thus, we can conclude that a filter ability to preserve edges and details is of prime importance for such "heterogeneous" classes. It can be also expected that the use of texture features for such classes can improve probability of their correct classification. Note that, for other classes, image pre-filtering also indirectly incorporates spatial information to classification by taking into account neighbouring pixel values at denoising stage to "correct" a given pixel value.

Let us now present examples of classification. Fig. 11,a, and 11,b illustrate classification results for noisy images (σ^2=100) for both classifiers. There are quite many pixel-wise misclassifications due to influence of noise, especially for the SVM classifier. Even the water surface is classified with misclassifications. In turn, Figures 11,c and 11,d present

Image	σ^2	Classifier	P_{corr1}	P_{corr2}	P_{corr3}	P_{corr4}	P_{corr5}
Noisy	49	RBF NN	0.717	0.909	0.987	0.718	0.805
Noisy	49	SVM	0.612	0.939	0.930	0.650	0.785
Filtered 2.1	49	RBF NN	0.814	0.991	0.987	0.715	0.830
Filtered 2.1	49	SVM	0.770	0.996	0.971	0.655	0.812
Filtered2.5	49	RBF NN	0.827	0.994	0.987	0.714	0.833
Filtered 2.5	49	SVM	0.803	0.998	0.974	0.657	0.818
Filtered 3D	49	RBF NN	0.839	0.997	0.987	0.720	0.860
Filtered 3D	49	SVM	0.882	0.998	0.986	0.682	0.862
Noisy	100	RBF NN	0.649	0.790	0.984	0.718	0.776
Noisy	100	SVM	0.530	0.826	0.834	0.634	0.745
Filtered 2.1	100	RBF NN	0.811	0.983	0.986	0.718	0.819
Filtered 2.1	100	SVM	0.728	0.994	0.966	0.653	0.797
Filtered2.5	100	RBF NN	0.834	0.991	0.985	0.717	0.830
Filtered 2.5	100	SVM	0.776	0.998	0.969	0.658	0.805
Filtered 3D	100	RBF NN	0.853	0.996	0.984	0.719	0.862
Filtered 3D	100	SVM	0.888	0.998	0.985	0.687	0.858

Table 2. Classification results for particular classes of original and filtered images

classification results for the three-channel image processed by the 3D DCT-based filter. It is clearly seen that quite many misclassifications have been corrected and the objects of certain classes have become compact. Comparison of the classification results in Figures 11,c and 11,d to the data in Figures 11,a and 11,b clearly demonstrate expedience of using RS image pre-filtering before classification if noise is intensive.

Let us give one more example for multichannel radar imaging. Fig. 12,a shows a three-channel radar image (in monochrome representation composed of HH Ka-band, VV Ka-band, and HH X-band SLAR images. The result of its component-wise processing by the modified sigma filter is presented in Fig. 12,b. Noise is suppressed but the edges are smeared due to residual errors of image co-registration and low contrasts of edges.

Considerably better edge/detail preservation is provided by the vector filter (Kurekin et al., 1997) that, in fact, sharpens edges if their misalignment in component images is detected (see Fig. 12,c). Finally, the result of bare soil areas detection (pixels are shown by white) by trained RBF NN applied to filtered data is depicted in Fig. 12,d. Since we had topology map for this region, probability of correct detection has been calculated and it was over 0.93. Classification results from original co-registered images were considerably less accurate.

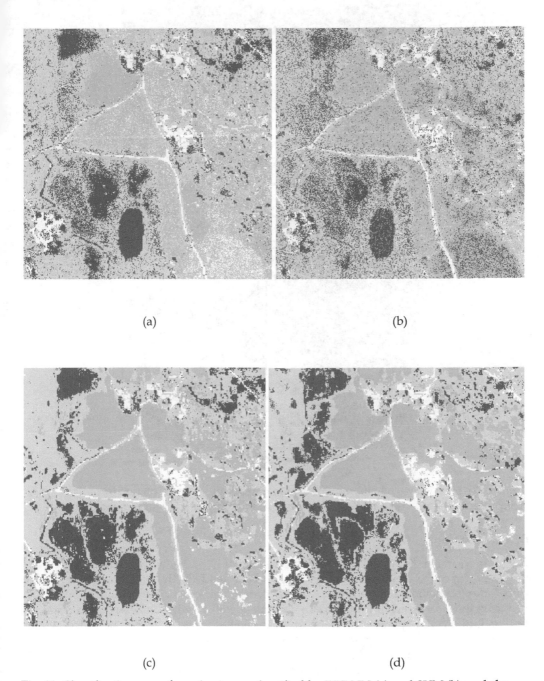

(a) (b)

(c) (d)

Fig. 11. Classification maps for noisy image classified by RBF NN (a) and SVM (b) and the image pre-processed by the 3D DCT filter classified by RBF NN (c) and SVM (d)

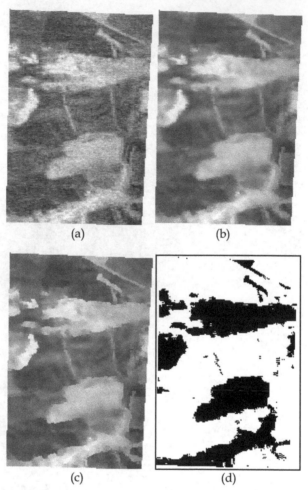

Fig. 12. Original three-channel radar image in monochrome representation (a), output for component-wise processing (b), output for vector filtering (c), classification map (d)

5. Conclusions

It is demonstrated that in most modern applications of multichannel RS noise characteristics deviate from conventional assumption to be additive and i.i.d. Thus, filtering techniques are to be adapted to more sophisticated real-life models. This especially relates to multichannel radar imaging for which it is possible to gain considerably higher efficiency of denoising by taking into account spatial correlation of noise and sufficient correlation of information in component images. New approaches that take into account aforementioned properties are proposed and tested for real life data. It is also shown that filtering is expedient for RS images contaminated by considerably less intensive noise than in radar imaging. Even if noise is practically not seen (noticeable by visual inspection) in original images, its removal by efficient filters can lead to increase of data classification accuracy.

6. References

Abramov, S., Zabrodina, V., Lukin, V., Vozel, B., Chehdi, K., & Astola, J. (2011). Methods for Blind Estimation of the Variance of Mixed Noise and Their Performance Analysis, In: *Numerical Analysis – Theory and Applications*, Jan Awrejcewicz (Ed.), InTech, ISBN 978-953-307-389-7, Retrieved from <http://www.intechopen.com/articles/show/title/methods-for-blind-estimation-of-the-variance-of-mixed-noise-and-their-performance-analysis>

Aiazzi, B., Alparone, L., Barducci, A., Baronti, S., Marcoinni, P., Pippi, I., & Selva, M. (2006). Noise modelling and estimation of hyperspectral data from airborne imaging spectrometers. *Annals of Geophysics*, Vol. 49, No. 1, February 2006

Ainsworth, T., Lee, J.-S., & Chang, L.W. (2007). Classification Comparisons between Dual-Pol and Quad-Pol SAR Imagery, *Proceedings of IGARSS*, pp. 164-167

Alberga, V., Satalino, G., & Staykova, D. (2008). Comparison of Polarimetric SAR Observables in Terms of Classification Performance. *International Journal of Remote Sensing*, Vol. 29, Issue 14, (July 2008), pp. 4129-4150

Amato, U., Cavalli, R.M., Palombo, A., Pignatti, S., & Santini, F. (2009). Experimental approach to the selection of the components in the minimum noise fraction, *IEEE Transactions on Geoscience and Remote Sensing*, Vol. 47, No 1, pp. 153-160

Astola, J., Haavisto, P. & Neuvo, Y. (1990) Vector Median Filters, Proc. IEEE, 1990, Vol. 78, pp. 678-689

Barducci, A., Guzzi, D., Marcoionni, P., & Pippi, I. (2005). CHRIS-Proba performance evaluation: signal-to-noise ratio, instrument efficiency and data quality from acquisitions over San Rossore (Italy) test site, *Proceedings of the 3-rd ESA CHRIS/Proba Workshop*, Italy, March 2005

Benedetto, J.J., Czaja, W., Ehler, M., Flake, C., & Hirn, M. (2010). Wavelet packets for multi- and hyperspectral imagery, *Proceedings of SPIE Conference on Wavelet Applications in Industrial Processing XIII*, SPIE Vol. 7535

Berge, A. & Solberg, A. (2004). A Comparison of Methods for Improving Classification of Hyperspectral Data, *Proceedings of IGARSS*, Vol. 2, pp. 945-948

Bose, N.K. & Liang, P. (1996). *Neural network fundamentals with graphs, algorithms and applications*, McGraw Hill

Chatterjee, P. & Milanfar, P. (2010). Is Denoising Dead? *IEEE Transactions on Image Processing*, Vol. 19, No 4, (April 2010), pp. 895-911

Chatterjee, P. & Milanfar, P. (2011). Practical Bounds on Image Denoising: From Estimation to Information. *IEEE Transactions on Image Processing*, , Vol. 20, No 5, (2011), pp. 221-1233

Chein-I Chang (Ed.) (2007). *Hyperspectral Data Exploitation: Theory and Applications*, Wiley-Interscience

Chen, G. & Qian, S. (2011). Denoising of Hyperspectral Imagery Using Principal Component Analysis and Wavelet Shrinkage. *IEEE Transactions on Geoscience and Remote Sensing*, Vol. 49, pp. 973-980

Christophe, E., Leger, D., & Mailhes, C. (2005). Quality criteria benchmark for hyperspectral imagery. *IEEE Transactions on Geoscience and Remote Sensing*, No. 43(9), pp. 2103-2114.

Cristianini, N. & Shawe-Taylor, J. (2000). *An Introduction to Support Vector Machines and Other Kernel-based Learning Methods*, Cambridge University Press

Curran, P.J. & Dungan, J., L. (1989). Estimation of signal-to-noise: a new procedure applied to AVIRIS data. *IEEE Transactions on Geoscience and Remote Sensing*, Vol. 27, pp. 20 – 628.

Dabov, K., Foi,A., & Egiazarian, K. (2007). Video Denoising by Sparse 3D-Transform Domain Collaborative Filtering, *Proceedings of EUSIPCO*, 2007

De Backer, S., Pizurica,A., Huysmans, B., Philips, W., & Scheunders, P. (2008). Denoising of multicomponent images using wavelet least squares estimators. *Image and Vision Computing*, Vol. 26, No 7, pp. 1038-1051

Deledalle, C.-A., Tupin, F., & Denis, L. (2011). Patch Similarity under Non Gaussian Noise, *Proceedings of ICIP*, 2011

Demir, B., Erturk, S., & Gullu, K. (2011). Hyperspectral Image Classification Using Denoising of Intrinsic Mode Functions. *IEEE Geoscience and Remote Sensing Letters*, Vol. 8, No 2, pp. 220-224.

Elad, M. (2010). *Sparse and Redundant Representations. From Theory to Applications in Signal and Image Processing*, Springer Science+Business Media, LLC

Ferro-Famil, L. & Pottier, E. (2001). Multi-frequency polarimetric SAR data classification, *Annals Of Telecommunications*, Vol. 56, No 9-10, pp. 510-522

Fevralev, D., Lukin, V., Ponomarenko, N., Vozel, B., Chehdi, K., Kurekin, A., & Shark, L. (2010). Classification of filtered multichannel images, *Proceedings of SPIE/EUROPTO on Satellite Remote Sensing*, Toulouse, France, September 2010

Fevralev, D., Ponomarenko, N., Lukin, V., Abramov, S., Egiazarian, K., & Astola, J. (2011). Efficiency analysis of color image filtering. *EURASIP Journal on Advances in Signal Processing*, 2011:41

Foi, A., Dabov, K., Katkovnik, V., & Egiazarian, K. (2007). Image denoising by sparse 3-D transform-domain collaborative filtering. *IEEE Transactions on Image Processing*, Vol. 6, No 8, (2007), pp. 2080-2095

Gungor, O. & Shan, J. (2006). An optimal fusion approach for optical and SAR images, *Proceedings of ISPRS Commission VII Mid-term Symposium „Remote Sensing: from Pixels to Processes"*, Netherlands, May 2006, pp. 111-116

Jeon, B. & Landgrebe, D.A. (1999). Partially supervised classification using weighted unsupervised clustering. *IEEE Transactions on Geoscience and Remote Sensing*, Vol. 37, No 2, pp. 1073–1079

Kerekes, J.P. & Baum, J.E. (2003). Hyperspectral Imaging System Modeling. *Lincoln Laboratory Journal*, Vol. 14, No 1, pp. 117-130

Kervrann, C. & Boulanger, J. (2008). Local adaptivity to variable smoothness for exemplar-based image regularization and representation. *International Journal of Computer Vision*, Vol. 79, No 1, (2008), pp. 45-69

Kulemin, G.P., Zelensky, A.A., & Astola, J.T. (2004). Methods and Algorithms for Pre-processing and Classification of Multichannel Radar Remote Sensing Images, *TICSP Series*, No. 28, ISBN 952-15-1293-8, Finland, TTY Monistamo

Kurekin, A.A., Lukin, V.V., Zelensky, A.A., Ponomarenko, N.N., Astola, J.T., & Saarinen, K.P. (1997). Adaptive Nonlinear Vector Filtering of Multichannel Radar Images, *Proceedings of SPIE Conference on Multispectral Imaging for Terrestrial Applications II*, San Diego, CA, USA, SPIE Vol. 3119, pp. 25-36

Kurekin, A.A., Lukin, V.V., Zelensky, A.A., Koivisto, P.T., Astola, J.T., & Saarinen, K.P. (1999). Comparison of component and vector filter performance with application to multichannel and color image processing, *Proceedings of the IEEE-EURASIP Workshop on Nonlinear Signal and Image Processing*, Antalya, Turkey, June 1999, No. 1, pp. 38-42

Landgrebe, D. (2002). Hyperspectral image data analysis as a high dimensional signal problem. *IEEE Signal Processing Magazine*, No. 19, pp. 17-28

Lee, J.-S. (1983). Digital Image Smoothing and the Sigma Filter. *Comp. Vis. Graph. Image Process.*, No. 24, (1983), pp. 255-269

Lukin, V., Tsymbal, O., Vozel, B., & Chehdi, K. (2006). Processing multichannel radar images by modified vector sigma filter for edge detection enhancement, *Proceedings of ICASSP*, Vol II, pp 833-836

Lukin, V., Fevralev, D., Ponomarenko, N., Abramov, S., Pogrebnyak, O., Egiazarian, K., & Astola, J. (2010a). Discrete cosine transform-based local adaptive filtering of images corrupted by nonstationary noise. *Electronic Imaging Journal*, Vol. 19(2), No. 1, (April-June 2010)

Lukin, V., Ponomarenko, N., Zriakhov, M., Kaarna, A., & Astola, J. (2010b). An Automatic Approach to Lossy Compression of AVIRIS Hyperspectral Data. *Telecommunications and Radio Engineering*, Vol. 69(6), (2010), pp. 537-563.

Lukin, V., Abramov, S., Ponomarenko, N., Egiazarian, K., & Astola, J. (2011a). Image Filtering: Potential Efficiency and Current Problems, *Proceedings of ICASSP*, 2011, pp. 1433-1436

Lukin, V., Abramov, S., Ponomarenko, N., Uss, M, Zriakhov, M., Vozel, B., Chehdi, K., Astola, J. (2011b). Methods and automatic procedures for processing images based on blind evaluation of noise type and characteristics. *SPIE Journal on Advances in Remote Sensing*, 2011, DOI: 10.1117/1.3539768

Makitalo, M., Foi, A., Fevralev, D., & Lukin, V. (2010). Denoising of single-look SAR images based on variance stabilization and non-local filters, *CD-ROM Proceedings of MMET*, Kiev, Ukraine, September 2010

Melgani, F. & Bruzzone, L. (2004). Classification of Hyperspectral Remote Sensing Images with Support Vector Machines, *IEEE Transactions on Geoscience and Remote Sensing*, Vol. 42, No 8, pp. 1778-1790

Murtagh, F., Starck, J.L., & Bijaoui, A. (1995). Image restoration with noise suppression using a multiresolution support, *Astron. Astrophys. Suppl. Ser*, 112, pp. 179-189

Oliver, C. & Quegan, S. (2004). *Understanding Synthetic Aperture Radar Images*, SciTech Publishing

Phillips, R.D., Blinn, C.E., Watson, L.T., & Wynne, R.H. (2009). An Adaptive Noise-Filtering Algorithm for AVIRIS Data With Implications for Classification Accuracy. *IEEE Transactions of GRS*, Vol. 47, No 9, (2009), pp. 3168-3179

Pizurica, A. & Philips, W. (2006). Estimating the probability of the presence of a signal of interest in multiresolution single- and multiband image denoising. *IEEE Transactions on Image Processing*, Vol. 15, No 3, pp. 654-665

Plataniotis, K.N. & Venetsanopoulos, A.N. (2000). *Color Image Processing and Applications*, Springer-Verlag, NY

Plaza, J., Plaza, A., Perez, R., & Martinez, P. (2008). Parallel classification of hyperspectral images using neural networks, *Comput. Intel. for Remote Sensing, Springer SCI*, Vol. 133, pp. 193-216

Ponomarenko, N., Lukin, V., Zriakhov, M., & Kaarna, A. (2006). Preliminary automatic analysis of characteristics of hyperspectral AVIRIS images, *Proceedings of MMET*, Kharkov, Ukraine, pp. 158-160

Ponomarenko, N., Silvestri, F., Egiazarian, K., Carli, M., Astola, J., & Lukin, V. (2007). On between-coefficient contrast masking of DCT basis functions, *CD-ROM Proceedings of the Third International Workshop on Video Processing and Quality Metrics*, USA, 2007

Ponomarenko, N., Lukin, V., Egiazarian, K., & Astola, J. (2008a). Adaptive DCT-based filtering of images corrupted by spatially correlated noise, *Proc. SPIE Conference Image Processing: Algorithms and Systems VI*, 2008, Vol. 6812

Ponomarenko, N., Lukin, V., Zelensky, A., Koivisto, P., & Egiazarian, K. (2008b). 3D DCT Based Filtering of Color and Multichannel Images, *Telecommunications and Radio Engineering*, No. 67, (2008), pp. 1369-1392

Ponomarenko, N., Lukin, V., Egiazarian, K., & Astola, J. (2010). A method for blind estimation of spatially correlated noise characteristics, *Proceedings of SPIE Conference Image Processing: Algorithms and Systems VII*, San Jose, USA, 2010, Vol. 7532

Ponomarenko, N., Lukin, V., & Egiazarian, K. (2011). HVS-Metric-Based Performance Analysis Of Image Denoising Algorithms, *Proceedings of EUVIP*, Paris, France, 2011

Popov, M.A., Stankevich, S.A., Lischenko, L.P., Lukin, V.V., & Ponomarenko, N.N. (2011). Processing of Hyperspectral Imagery for Contamination Detection in Urban Areas. *NATO Science for Peace and Security Series C: Environmental Security*, pp. 147-156.

Rellier, G., Descombes, X., Falzon, F., & Zerubia, J. (2004). Texture Feature Analysis using a Gauss-Markov Model in Hyperspectral Image Classification, *IEEE Transactions in Geoscience and Remote Sensing*, Vol. 42, No 7, pp. 1543-1551

Renard, N., Bourennane, S., & Blanc-Talon, J. (2006). Multiway Filtering Applied on Hyperspectral Images, *Proceedings of ACIVS, Springer LNCS*, Vol. 4179, pp. 127-137

Renard, N., Bourennane, S., & Blanc-Talon, J. (2008). Denoising and Dimensionality Reduction Using Multilinear Tools for Hyperspectral Images. *IEEE Geoscience and Remote Sensing Letters*, Vol. 5, No 2, pp. 138-142

Schölkopf, B., Burges, J.C., & Smola, A.J. (1999). *Advances in Kernel Methods: Support Vector Learning*, MIT Press, Cambridge, MA.

Schowengerdt, R.A. (2007). *Remote Sensing: Models and Methods for Image Processing*, Academic Press

Tsymbal, O.V., Lukin, V.V., Ponomarenko, N.N., Zelensky, A.A., Egiazarian, K.O., & Astola, J.T. (2005). Three-state Locally Adaptive Texture Preserving Filter for Radar and Optical Image Processing. *EURASIP Journal on Applied Signal Processing*, No. 8, (May 2005), pp. 1185-1204

Uss, M., Vozel, B., Lukin, V., & Chehdi, K. (2011a). Local Signal-Dependent Noise Variance Estimation from Hyperspectral Textural Images. *IEEE Journal of Selected Topics in Signal Processing*, Vol. 5, No. 2, DOI: 10.1109/JSTSP.2010.2104312

Uss, M., Vozel, B., Lukin, V., & Chehdi, K. (2011b). Potential MSE of color image local filtering in component-wise and vector cases, *Proceedings of CADSM*, Ukraine, February 2011, pp. 91-101

Yli-Harja, O. & Shmulevich, I. (1999). Correcting Misclassifications in Hyperspectral Image Data Using a Nonlinear Graph-based Estimation Technique, *International Symposium on Nonlinear Theory and its Applications*, pp. 259-262

Zabala, A., Pons, X., Diaz-Delgado, R., Garcia, F., Auli-Llinas, F., & Serra-Sagrista, J. (2006). Effects of JPEG and JPEG2000 lossy compression on remote sensing image classification for mapping crops and forest areas, *Proceedings of IGARSS*, pp. 790-793

Zelensky, A., Kulemin, G., Kurekin, A., & Lukin, V. (2002). Modified Vector Sigma Filter for the Processing of Multichannel Radar Images and Increasing Reliability of Its Interpretation. *Telecommunication and Radioengineering*, Vol. 58, No. 1-2, pp.100-113

Zhang, H., Peng, H., Fairchild, M.D., & Montag, E.D. (2008). Hyperspectral Image Visualization based on Human Visual Model, *Proceedings of SPIE Conference on Human Vision and Electronic Imaging XIII*, SPIE Vol. 6806

4

Fusion of Optical and Thermal Imagery and LiDAR Data for Application to 3-D Urban Environment and Structure Monitoring

Anna Brook[1], Marijke Vandewal[1] and Eyal Ben-Dor[2]
[1]Royal Military Academy, CISS Department, Brussels
[2]Remote Sensing Laboratory, Department of Geography and Environment,
Tel-Aviv University, Tel-Aviv
[1]Belgium
[2]Israel

1. Introduction

For many years, panchromatic aerial photographs have been the main source of remote sensing data for detailed inventories of urban areas. Traditionally, building extraction relies mainly on manual photo-interpretation which is an expensive process, especially when a large amount of data must be processed (Ameri, 2000). The characterization of a given object bases on its visible information, such as: shape (external form, outline, or configuration), size, patterns (spatial arrangement of an object into distinctive forms), shadow (indicates the outlines, length, and is useful to measure height, or slopes of the terrain), tone (color or brightness of an object, smoothness of the surface, etc.) (Ridd 1995). Automated assessment of urban surface characteristics has been investigated due to the high costs of visual interpretation. Most of those studies used multispectral satellite imagery of medium to low spatial resolution (Landsat-TM, SPOT-HRV, IRS-LISS, ALI and CHRIS-PROBA) and were based on common image-analysis techniques (e.g. maximum likelihood (ML) classification, principal components analysis (PCA) or spectral indices (Richards and Jia 1999)). The problems of limited spatial resolution over urban areas have been overcome with the wider availability of space-borne systems, which characterized by large swath and high spatial and temporal resolutions (e.g. Worl-View2). However, the limits on spectral information of non-vegetative material render their exact identification difficult. In this regard, the hyperspectral remote sensing (HRS) technology, using data from airborne sensors (e.g. AVIRIS, GER, DAIS, HyMap, AISA-Dual), has opened up a new frontier for surface differentiation of homogeneous material based on spectral characteristics (Heiden et al. 2007). This capability also offers the potential to extract quantitative information on biochemical, geochemical and chemical parameters of the targets in question (Roessner et al. 1998).

The most common approach to characterizing urban environments from remote sensing imagery is land-use classification, i.e. assigning all pixels in the image to mutually exclusive classes, such as residential, industrial, recreational, etc. (Ridd 1995, Price 1998). In contrast, mapping the urban environment in terms of its physical components preserves the heterogeneity of urban land cover better than traditional land-use classification (Jensen & Cowen, 1999), characterizes urban land cover independent of analyst-imposed definitions and more accurately captures changes with time (Rashed et al. 2001).

Hyperspectral thermal infrared (TIR) remote sensing has rapidly advanced with the development of airborne systems and follows years of laboratory studies (Hunt & Vincent 1968, Conel 1969, Vincent & Thomson 1972, Logan et al. 1975, Salisbury et al. 1987). The radiance emitted from a surface in thermal infrared (4-13μm) is a function of its temperature and emissivity. Emittance and reflectance are complex processes that depend not only on the absorption coefficient of materials but also on their reflective index, physical state and temperature. Most urban built environment studies are taking into account both temperature and emissivity variations, since these relate to the targets identification, mapping and monitoring and provide a mean for practical application.

The hyperspectral thermal imagery provides the ability for mapping and monitoring temperatures related to the man-made materials. The urban heat island (UHI) has been one of the most studied and the best-known phenomena of urban climate investigated by thermal imagery (Carlson et al., 1981; Vukovich, 1983; Kidder & Wu, 1987; Roth et al., 1989; Nichol, 1996). The preliminary studies have reported similarities between spatial patterns of air temperature and remotely sensed surface temperature (Henry et al., 1989; Nichol 1994), whereas progress studies suggest significant differences, including the time of day and season of maximum UHI development and the relationship between land use and UHI intensity (Roth et al., 1989). The recent high-resolution airborne systems determine the thermal performance of the building that can be used to identify heating and cooling loss due to poor construction, missing or inadequate insulation and moisture intrusion.

The spectral (reflective and thermal) characteristics of the urban surfaces are known to be rather complex as they are composed of many materials. Given the high degree of spatial and spectral heterogeneity within various artificial and natural land cover categories, the application of remote sensing technology to mapping built urban environments requires specific attention to both 3-D and spectral domains (Segl et al. 2003). Segl confirms that profiling hyperspectral TIR can successfully identify and discriminate a variety of silicates and carbonates, as well as variations in the chemistry of some silicates. The integration of VNIR-SWIR and TIR results can provide useful information to remove possible ambiguous interpretations in unmixed sub-pixel surfaces and materials. The image interpretation is based on the thematic categories (Roessner et al. 2001), which are defined by the rules of urban mapping and land-uses.

The ultimate aim in photogrammetry in generating an urban landscape model is to show the objects in an urban area in 3-D (Juan et al. 2007). As the most permanent features in the urban environment, an accurate extraction of buildings and roads is significant for urban planning and cartographic mapping. Acquisition and integration of data for the built urban environment has always been a challenge due to the high cost and heterogeneous nature of the data sets (Wang 2008). Thus, over the last few years, LiDAR (LIght Detection And Ranging) has been widely applied in the field of photogrammetry and urban 3-D analysis (Tao 2001, Zhou 2004). Airborne LiDAR technique provides geo-referenced 3-D dense points ("cloud") measured roughly perpendicular to the direction of flight over a reflective surface on the ground. This system integrates three basic data-collection tools: a laser scanner, a global positioning system (GPS) and an inertial measuring unit (IMU). The position and altitude of it determined by GPS/INS, therefore, the raw data are collected in the GPS reference system WGS 84.

Generally, 3-D urban built environment models are created using CAD (computer-aided design) tools. There have been many successful projects which have produced detailed and realistic 3-D models for a diverse range of cities (Dodge et al. 1998, Bulmer 2001, Jepson et al. 2001). These city models were created with accurate building models compiled with orthophotographs and exhibited an impressive, realistic urban environment (Chan et al. 1998). However, the creation of 3-D city models using CAD tools and orthophotographs faces some challenges: it is time-consuming and expensive.

The analysis of InSAR (Interferometric Synthetic Aperture Radar) and SAR (Synthetic Aperture Radar) data for urban built targets has several important benefits, such as the ability to adopt numerical tools, and the ability to provide results resembling the real-world situation. In addition, a relation can be found between target geometry and the measured scattering, and according to target-scattering properties, height-retrieval algorithms can be developed. The limitation of this method is that the targets in urban models have to be as detailed as possible; otherwise the results obtained in the modeled environment will be not reliable (Margarit et al. 2007).

The use of 3-D high-spatial-resolution applications in urban built environments is a mainstay of architecture and engineering practice. However, engineering practices are increasingly incorporating different data sets and alternative dissemination systems. Understanding, modeling and forecasting the trends in urban environments are important to recognize and assess the impact of urbanization for resource managers and urban planners. Many applications are suitable sources of reliable information on the multiple facets of the urban environment (Jensen & Cowen 1999, Donnay et al. 2001, Herold et al. 2003). These models have provided simulations of urban dynamics and an understanding of the patterns and processes associated with urbanization (Herold et al. 2005). However, the complexity of urban systems makes it difficult to adequately address changes using a single data type or analysis approach (Allen & Lu 2003).

This chapter presents techniques for data fusion and data registration. The ability to include an accurate and realistic 3-D position, quantitative spectral information, thermal properties and temporal changes provides a near-real-time monitoring system for photogrammetric and urban planning purposes. The method is focusing on registration of multi-sensor and multi-temporal information for 3-D urban environment monitoring applications. Generally, data registration is a critical pre-processing procedure in all remote-sensing applications that utilizes multiple sensors inputs, including multi-sensor data fusion, temporal change detection, and data mosaicking. The main objective of this research is a fully controlled, near-real-time, natural and realistic monitoring system for an urban environment. This task led us first to combine the image-processing and map-matching procedures, and then to incorporate remote sensing and GIS tools into an integrative method for data fusion and registration. To support this new data model, traditional spatial databases were extended to support 5-D data.

This chapter is organized as follows. Section 2 describes the materials and methods, which are implemented in the 3-D urban environment model presented in Section 3. Section 4 addresses to the generic 3-D urban application, which involves data fusion and contextual information of the environment.

2. Materials and methods

2.1 Study area

Two separate datasets were utilized in this study. The first dataset was acquired over the suburban Mediterranean area on 10 Oct 2006 at 03h37 UTC and at 11h20 UTC. This area combines natural and engineered terrains (average elevation of 560m above sea level), a hill in the north of the studied polygon area and a valley in the center. The entire scene consists of rows of terraced houses located at the center of the image. The neighborhood consists of cottage houses (two and three floors) with tile roofs, flat white-colored concrete roofs and balconies, asphalt roads and parking lots, planted and natural vegetation, gravel paths and bare brown forest soil. The height of large buildings ranges from 8 to 16 m. A group of tall pine trees with various heights and shapes are located on the streets and the Mediterranean forest can be found in the corner of the scene.

The second dataset was acquired over urban settlement, on 15 Aug 2007 at 02h54 UTC and at 12h30 UTC. This area combines natural, agriculture and engineered terrains (average elevation of 30m above sea level). The urban settlement consists of houses (two and three floors) and public buildings (schools and municipalities buildings) with flat concrete, asphalt or whitewash roofing, asphalt roads and parking lots, planted and natural vegetation, gravel paths, bare brown reddish Mediterranean and agriculture soils, greenhouses and whitewash henhouse roofing. The height of large buildings ranges from 3 to 21 m.

2.2 Data-acquisition systems

The research combines airborne and ground data collected from different platforms and different operated systems. The collected imagery data were validated and compared to the ground truth in situ measurements collected during the campaigns.

The first airborne platform carries AISA-Dual hyperspectral system. The airborne imaging spectrometer AISA-Dual (Specim Ltd.) is a dual hyperspectral pushbroom system, which combines the Aisa EAGLE (VNIR region) and Aisa HAWK (SWIR region) sensors. For the selected campaigns, the sensor simultaneously acquired images in 198 contiguous spectral bands, covering the 0.4 to 2.5 µm spectral region with bandwidths of ~10 nm for Aisa EAGLE and ~5 nm for Aisa HAWK. The sensor altitude was 10,000 ft, providing a 1.6 m spatial resolution for 286 pixels in the cross-track direction. A standard AISA-Dual data set is a 3-D data cube in a non-earth coordinate system (raw matrix geometry).

The second airborne platform carries hyperspectral TIR system, which is a line-scanner with 28 spectral bands in the thermal ranges 3-5 µm and 8-13 µm. It has 328 pixels in the cross-track direction and hundreds of pixels in the along-track direction with a spatial resolution of 1.4m.

The third airborne platform carries the LiDAR system. This system operates at 1500 nm wavelength with a 165 kHz laser repetition rate and 100 Hz scanning rate and provides a spatial/footprint resolution of 0.5 m and an accuracy of 0.1 m. The scanner has a multi-pulse system that could record up to five different returns, but in this study, only the first return was recorded and analyzed.

The ground spectral camera HS (Specim Ltd.) is a pushbroom scan camera that integrate ImSpector imaging spectrograph and an area monochrome camera. The camera's sensitive high speed interlaced CCD (Charge-Coupled Device) detector simultaneously acquires images in 850 contiguous spectral bands and covers the 0.4 to 1 µm spectral region with bandwidths of 2.8 nm. The spatial resolution is 1600 pixels in the cross-track direction, and the frame rate is 33 fps with adjustable spectral sampling.

The ground truth reflectance data were measured for the calibration/validation targets by the ASD "FieldSpec Pro" (ASD.Inc, Boulder, CO) VNIR-SWIR spectrometer. Internally averaged scans were 100 ms each. The wavelength-dependent signal-to-noise ratio (S/N) is estimated by taking repeat measurements of a Spectralon white-reference panel over a 10-min interval and analyzing the spectral variation across this period. For each sample, three spectral replicates were acquired and the average was used as the representative spectrum. The ground truth thermal data were collected by a thermometer and thermocouples installed within calibration/validation targets (water bodies) and a thermal radiometer infrared camera (FLIR Systems, Inc.).

2.3 Data processing

This research integrates multi-sensor (airborne sensor, ground camera and field devises) and multi-temporal information into fully operational monitoring application. The aim of this sub-paragraph is to present several techniques for imagery and LiDAR data processing.

The classification approaches for airborne and ground hyperspectral imagery are firstly presented. The radiance measured by these sensors strongly depends on the atmospheric conditions, which might bias the results of material identification/classification algorithms that rely on hyperspectral image data. The desire to relate imagery data to intrinsic surface properties has led to the development of atmospheric correction algorithms that attempt to recover surface reflectance or emission from at-sensor radiance. Secondly, the LiDAR data are processed by applying the surface-based clustering methods.

2.3.1 Hyperspectral airborne and ground imagery

Accurate spectral reflectance information is a key factor in retrieving correct thematic results. In general, the quality of HRS sensors varies from very high to moderate (and even very poor) in terms of signal-to-noise ratio, radiometric accuracy and sensor stability. Instability of the sensors' radiometric performance (stripes, saturation, etc.) might be caused by either known or unknown factors encountered during sensor transport, installation and/or even data acquisition. As part of data pre-processing, these distortions have to be assessed and quantified for each mission.

A full-chain atmospheric calibration SVC (supervised vicarious calibration) method (Brook & Ben-Dor 2011a) is applied to extract reflectance information from hyperspectral imagery. This method is based on a mission-by-mission approach, followed by a unique vicarious calibration site. In this study, the acquired AISA-Dual and HS images were subjected to the SVC method, which includes two radiometric recalibration techniques (F1 and F2) and two atmospheric correction approaches (F3 and F4). The atmospheric correction incorporate deshadow algorithm, which is applied o the map provided by the boresight ratio band (Brook & Ben-Dor 2011b).

The hyperspectral reflectance images are subjected to the data processing stage, which is operated in four steps (Figure 1). First step is a general coarse classification. Each "pure" pixel is assigned to a class in order to predefine the threshold of the probabilistic output of a support vector machine (SVM) algorithm, or remains unclassified (Villa et al., 2011). The unclassified pixels might associate with mixed spectra pixels, thus their classification is addressed at the third stage by the unmixing method in order to obtain the abundance fraction of each endmember class. Prior to this step, a second step is applied, where spectral data are reduced by the selected algorithm. The input variables in terms of absorption features can be reduced through a sequential forward selection (SFS) algorithm (Whitney, 1971). This method starts with the inclusion of feature sets one by one to minimize the prediction error of a linear regression model and focuses on conditional exclusion based on feature significance (Pudil et al., 1994). This step is proven to enhance overall performance of spectral models.

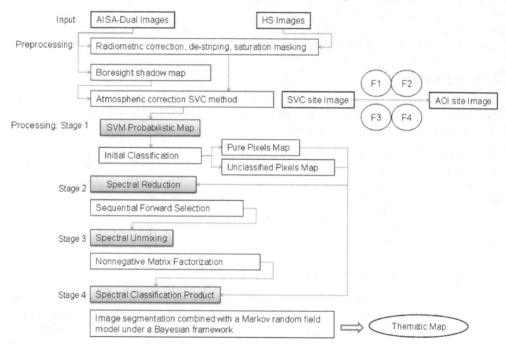

Fig. 1. Flow chart scheme of the classification approach for hyperspectral airborne and ground data

The nonnegative matrix factorization (NMF) was offered as an alternative method for linear unmixing (Lee et al., 2000). This algorithm search for the source and the transform by factorizing a matrix subject to positive constraints based on gradient optimization and Euclidean norm designation (Pauca et al, 2006; Robila & Maciak, 2006). We generated an algorithm that starts with the random linear transform to the nonnegative source data. The algorithm is continuously computing scalar factors that are chosen to produce the "best" intermediate source and transform. At each step of the algorithm the source and transform should remain positive. The final stage is a method for image segmentation combined with a Markov random field (MRF) model under a Bayesian framework (Yang & Jiang, 2003).

The validation of the thematic map is performed by comparing ground truth and image reflectance data of the selected targets. The ten well-known targets (areas of approximately 30-40 pixels) were spectrally measured (using ASD SpecPro) and documented. The overall accuracy for the Ma'alot Tarshiha images was 96.8 and for the Qalansawe images it was 97.4. The exact location of each target within the scenes was captured using aerial orthophoto and ground truth field survey. The confusion matrices (Tables 1 and 2) and ROC (receiver operating characteristic) curve (Table 3) were calculated by comparison between number of pixels in each class (concrete, asphalt, scuffed asphalt) and ground truth maps. The overall accuracy of both images stands in good agreement, thus it can be concluded that the suggested classification algorithm (Figure 1) performance is stable and accurate.

	Ground truth (%)		
Class	Concrete	Asphalt	Scuffed asphalt
Unclassified	0	0	0
Concrete	**96.2**	1.7	2.1
Asphalt	1.1	**98**	0.9
Scuffed Asphalt	2.8	0.2	**97**

Table 1. Confusion matrix of the Ma'alot Tarshiha image for selected classes (Correspondence accuracies are in bold.)

	Ground truth (%)		
Class	Concrete	Asphalt	Scuffed asphalt
Unclassified	0	0	0
Concrete	**96.3**	0.8	2.9
Asphalt	0	**98.4**	1.6
Scuffed Asphalt	4.5	0	**95.5**

Table 2. Confusion matrix of the Qalansawe image for selected classes (Correspondence accuracies are in bold.)

	Ma'alot image			Qalansawe image		
	Concrete	Asphalt	Scuffed asphalt	Concrete	Asphalt	Scuffed asphalt
DR	0.97	0.97	0.94	0.98	0.96	0.93
Area	0.99	0.99	0.96	0.99	0.98	0.95

Table 3. Detection rates (DR) of concrete, asphalt and scuffed asphalt for false alarm probability 0.1 according to ROC and area under the curve

2.3.2 Thermal airborne and ground imagery

Atmospheric correction is a key processing step for extracting information from thermal infrared imagery. The ground-leaving radiance combined with temperature/emissivity separation (TES) algorithms are generated and supplied to in-scene atmospheric compensation ISAC[1] (Young et al., 2002). This model requires only the calibrated, at-aperture radiance data to estimate the upwelling radiance and transmittance of the atmosphere. It is an effective atmospheric correction that produces spectra that compare favorably to the Planck function.

The ground truth must include several targets as water, sand or soil continuously measured by installed thermocouples. The generating atmospheric data cube may be used as an input to a temperature emissivity separation algorithm (normalized emissivity method). The proposed thermal classification method follows the same four stages of data processing (SVM's probabilistic map; data reduction; umnixing; classification) applied to the pre-processed emissivity imagery (Figure 2).

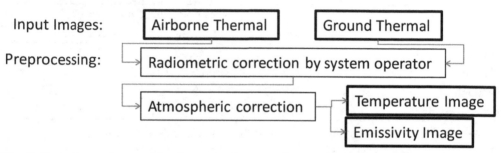

Fig. 2. Flow chart scheme of the thermal airborne and ground data preprocessing

From the physical definition, the spectral characteristics of urban materials in the reflective and thermal ranges are related. Segl (Segl et al., 2003) showed that materials with high albedos in the reflective range produce low albedos in the thermal range and vice versa, due to a better energy absorption in the reflective region. However, it is reported that bitumen roofing and asphalt pavement generate distinct spectral differences in the thermal wavelength range. The thermal measurements remain a compelling focus on a climate research in the built urban areas. However, the thermal airborne and ground imagery permit definition of UHI (for the ground surface) and resolve streets, roofs and walls. The successful numerical model of the urban areas is acquired during night-time conditions, when solar shading is absent and turbulent interactions are minimal.

The validation of the thematic map is performed by comparing ground truth and image emissivity data. The five targets (concrete, sand lot, bitumen, tile roof and polyethylene) were measured and documented. The resulting emissivity signatures are in good agreement with ground-truth data (two examples in Figure 3A and 3B). The results presented here confirm the robustness and stability of the suggested algorithm.

2.3.3 Airborne LiDAR data

LiDAR data provides precise information about the geometrical properties of the surfaces and can reflect the different shapes and formations in the complex urban environment. The point cloud (irregularly spaced points) was interpolated into the digital surface model (DSM) by applying the Kriging technique (Sacks et al. 1989). The Kriging model has its

---

[1] ISAC (in-scene atmospheric compensations) model is implemented in ENVI®

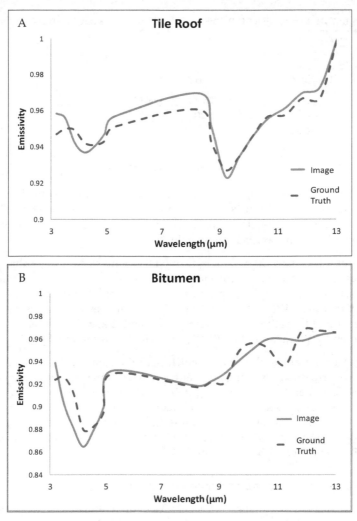

Fig. 3. Emissivity calculated from the thermal radiance. A is a tile roof and B is a bitumen
roof

origins in mining and geostatistical applications involving spatially and temporally
correlated data (Cressie 1993).

The surface analysis (Figure 4) is first represented as a DEM (digital elevation model) of the
scanned scene, where data are separated into on-terrain and off-terrain points (Masaharu
and Ohtsubo 2002). In this study, the Kriging Gaussian correlation function was utilized to
visualize and illustrate the edited DEM as a surface-response function. Note that the
interpolation converts irregularly spaced LiDAR data to a self-adaptive DSM.

The DTM (digital terrain model) was created by a morphological scale-opening filter, using
square structural elements (Rottensteiner et al., 2003). Then, according to the filter, the slope

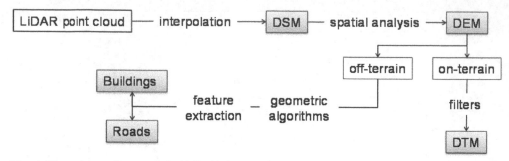

Fig. 4. Flow chart scheme of the LiDAR data surface analysis

map is estimated. The next stage is to fragment a surface model convolved with highly heterogeneous terrain slopes into subareas with fixed slope (Zhang et al., 2003; Shan & Sampath 2005). At this stage, the terrain is uniformly normalized and the separation between on- and off-terrain points is applicable.

The building boundary is determined by a modified convex hull algorithm (Jarvis 1973) which classifies the cluster data into boundary (contour/edge) and non-boundary (inter-shape) points (Jarvis 1977). Separating points located on buildings from those on trees and bushes, is a difficult task (Wang & Shan 2009). The common assumption is that the building outlines are separated from the trees in terms of size and shape. The dimensionality learning method, proposed by Wang and Shan (2009), is an efficient technique for this purpose.

In relatively flat urban areas, the roads, which have the same elevation (height) as a bare surface, can be extracted by arrangement examination. The simple geometric and topological relations between streets might be used to improve the consistency of road extraction. First, the DEM data are used to obtain candidate roads, sidewalks and parking lots. Then the road model is established, based on the continuous network of points which are used to extract information such as centerline, edge and width of the road (Akel et al. 2003; Hinz & Baumgartner 2003; Cloude et al., 2004).

2.4 Data registration: Automatic and manual approaches

The optical and thermal imagery and LiDAR data have fundamentally different characteristics. The LiDAR data (monochromatic NIR laser pulse) provides terrain characteristics; hence, optical imagery (radiation reflected back from the surface at many wavelengths) provides ability for in situ, easy, rapid and accurate assessment of many materials on a spatial/spectral/temporal domain, and thermal imagery determines temperatures and radiance signature of urban materials and land covers. Since all these datasets (Figure 5) are crucial for the assessment and classification of the urban area, a novel method for automatic registration and data fusion is needed.

Data fusion techniques combine data from multiple sensors and related information from associated databases. The integrated data set achieves higher accuracy and more specific inferences that might be obtained by the use of single sensor alone. In general, data registration is a critical preprocessing procedure in all remote-sensing applications that utilizes multiple sensor inputs, including multi-sensor data fusion, temporal change

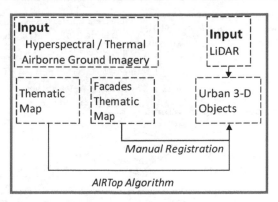

Fig. 5. Flow chart scheme of the input data and registration techniques

detection, and data mosaicking (Moigne et al., 2002). In manual registration, the selection of control points (CPs) is usually performed by a human operator. This has proven to be inaccurate, time-consuming, and unfeasible due to data complexity, which makes it cumbersome or even impossible for the human eye to discern the suitable CPs. Therefore, researchers have focused on automating feature detection to align two or more data sets with no need for human intervention.

The automatic registration of data sets has generated extensive research interest in the fields of computer vision, medical imaging and remote sensing. Comprehensive reviews have been published by Brown (1992) and Zitova and Flusser (2003). Many proposed schemes for automatic registration employ a multi-resolution process (Viola and Wells 1997, Wu & Chung 2004, Fan et al., 2005, Zavorin & Moigne 2005, Xu & Chen 2007).

The existing automatic data-registration techniques based on spatial information fall into two categories: intensity-based and feature-based (Zitova and Flusser 2003). The feature-based technique extracts salient structures from sensed and reference data sets by accurate feature detectin and by the overlap criterion. As the relevant objects of intereset (e.g., roofs) and lines (e.g., roads) are expected to be stable in time at a fixed position, the feature-based method is more suitable for multi-sensor and multi-data set fusion, change detection and mosaicking. The method generally consists of four steps (Jensen, 2004): (1) CP extraction; (2) transformation-model determination; (3) image transformation and resampling, and (4) assessment of registration accuracy. The first step is the most complex, and its success essentially determines registration accuracy. Thus, the detection method should be able to detect the same features in all projections and in different data, regardless of the particular image/sensor/data type deformation. Despite the achieved performance, the existing methods operate directly on gray intensity values and hence they are not suited for handling multi-sensor and multi-type data sets.

The suggested algorithm is an adapted version of the four stages AIRTop (Figure 6) algorithm (Brook & Ben-Dor 2011c). First, the significant features are extracted from all input data sets and converted to a vector format. Since the studied scene has a large area, regions of interest (ROI) with relatively large variations are selection. The idea of addressing the registration problem by applying a global-to-local level strategy (the whole image is now divided into regions of interest which are treated as an image) proves to be an elegant way

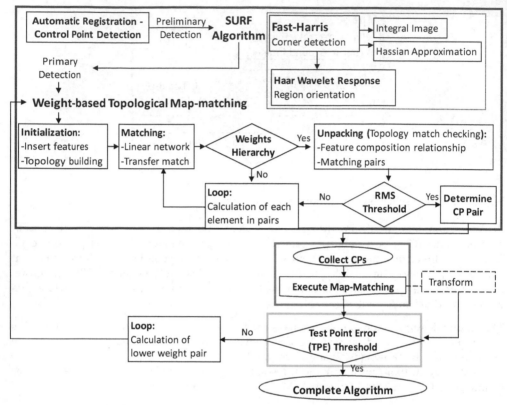

Fig. 6. A flow chart describing the registration algorithm. Blue box: topology map matching. Orange box: matching process. Green box: validation and accuracy.

of speeding up the whole process, while enhancing the accuracy of the registration procedure (Chantous et al. 2009). Thus, we expected this method to greatly reduce false alarms in the subsequent feature extraction and CP identification steps (Brook et al., 2011). To select the distinct areas in the vector data sets, a map of extracted features is divided into adjacent small blocks (10% × 10% of original image pixels with no overlap between blocks). Then, the significant CPs extraction has been performed by applying the SURF algorithm (Brown & Lowe, 2002). First the fast-Harris corners Detector (Lindeberg, 2004), which based on an integral image, was performed. The Hessian matrix is responsible for primary image rotation using principal points that identified as "interesting" potential CPs in the block. The local feature representing vector is made by combination of Haar wavelet response. The values of dominant directions are defined in relation to the principal point. As the number of interesting points tracked within the block is more than the predefined threshold, the block is selected and considered a suitable candidate for CPs detection.

The spatial distribution and relationship of these features are expressed by topology rules (one-to-one) and they are converted to potential CPs by determining a transformation model between sensed and reference data sets. The defined rules for a weight-based topological map-matching (tMM) algorithm manage (Velaga et al. 2009), transform and resample

features of the sensed goereferenced LiDAR data according to a non georeferenced imagery
in order to reserve original raw geometry, dimensionality and imagery matrices (imagery
pixels size and location).

In the proposed 3-D urban application, the manual registration is used to register facades
imagery and thematic mapping acquired by ground sensors and simplify buildings model
extracted from LiDAR. This method is executed by a human operator, who identifies a set of
corresponding CPs from the images and referenced control building model. Despite the fact
that manual registration has been proven inaccurate and time-consuming due to data
complexity, this method is still the most widely used technique. We found that for the
current data sets, manual registration is the easiest and most accurate solution.

3. 3-D urban environment model

The urban database-driven 3-D model represents a realistic illustration of the environment
that can be regularly updated with attribute details and sensor-based information. The
spatial data model is a hierarchical structure (Figure 7), consisting of elements, which make
up geometries, which in turn composes layers.

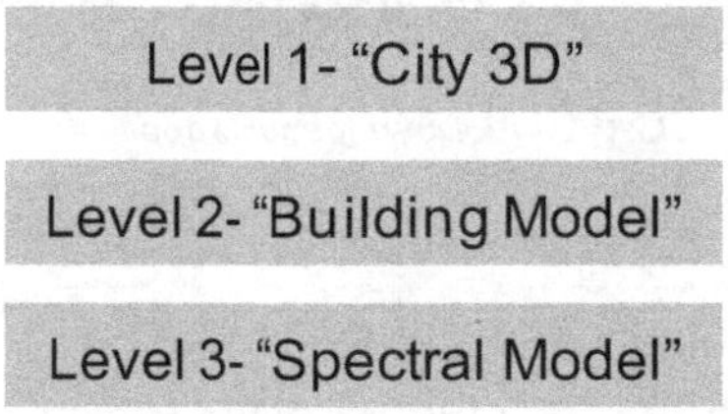

Level 1- "City 3D"

Level 2- "Building Model"

Level 3- "Spectral Model"

Fig. 7. The 3-D urban environment application's conceptual architecture

A fundamental demand in non-traditional, multi-sensors and multi-type applications is
spatial indexing. A spatial index, which is a logical index, provides a mechanism to limit
searches based on spatial criteria (such as intersection and containment). Due to the
variation of data formats and types, it is difficult to satisfy the frequent updating and
extension requirements for developing urban environments.

An R-tree index is implemented on spatial data by using Oracle's extensible indexing
framework (Song et al. 2009). This index approximates the geometry with a single rectangle
that minimally encloses the geometry (minimum bounding rectangle MBR). A bounding
volume is created around the 3-D object, which equals the bounding volume around the
solid. The index is helpful in conducting very fast searches and spatial analyses over large 3-
D scenes.

CityGML[2] is an application based on OGC's (open geospatial consortium) GML 3.1. This
application not only represents the graphical appearance but in particular, it takes care of
the semantic properties (Kolbe et al. 2005), such as the spectral/thematic properties, and
model evaluations. The main advantage is the ability to maintain different levels of detail
(Kolbe & Bacharach 2006). The underlying model differentiates three levels of detail, for
which objects become more detailed as the level incise.

---

[2] http://www.citygml.org/

The 3-D urban application is based on an integrated data set: spectral models, ground camera and airborne images, and LiDAR data. The system requirements are defined to include geo-spatial planning information and one-to-one topology. The concept architecture diagram is presented in Figure 4. As the model consist visualization and interactivity with maps and 3-D scenes, the interface includes 3-D interaction, 2-D vertical and horizontal interactions and browsers that contain spectral/thematic temporal information. The 3-D urban application provides services such as thematic mapping, and a complete quantitative review of the building and it's surrounding with respect to temporal monitoring. The design of the application shows the possibilities of delivering integrated information and thus holistic views of whole urban environments in a freeze-frame view of the spatiotemporal domain.

The self-sufficient/self-determining levels of the integrated information contribute different parts to this global urban environmental application. The first level (Figure 8), termed "City 3-D", supplies three different products: 1) integrated imagery and LiDAR data, 2) 3-D thematic map, and 3) 2-D thematic map (which includes 3-D analysis layers such as terrain properties, spatial analysis, etc.).

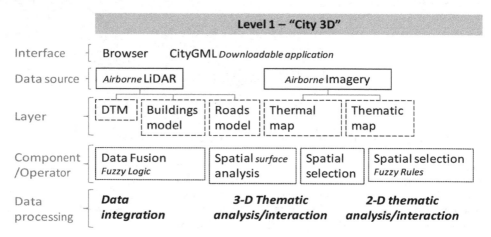

Fig. 8. The 3-D urban environment application – Level 1 (detailed architecture)

The second level, termed "Building Model" (Figure 9), focuses on a single building in 3-D and provides two additional products: 1) integrated imagery and building model extracted from LiDAR data set, and 2) 3-D thematic map for general materials classification, and quantitative thematic maps implemented by spectral models.

The most specific and localized level is the third level, termed "Spectral Model" (Figure 9). The area of interest in this level is a particular place (a patch) on the wall of the building in question. The spatial investigation at this level is a continuation of the previous level; yet, the data source consists of spectral models that are evaluated for spectral in-situ point measurements. This level does not provide any integrated and rectified information, but provides geo-referencing of the results of the spectral models in realistic 3-D scale. This level completes the database of the suggested 3-D urban environment application.

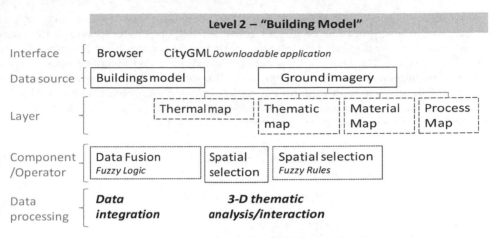

Fig. 9. The 3-D urban environment application – Level 2 (detailed architecture)

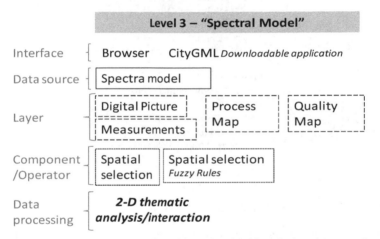

Fig. 10. The 3-D urban environment application – Level 3 (detailed architecture)

4. 3-D urban environment application

The 3-D monitoring built urban environment application, up to this point, employs single processing algorithms applied on imagery or LiDAR data, without taken into account contextual information. The data fusion application must provide fully integrated information, both of the classification products and the context within the scene. In the proposed application, a complete classification and identification task consist of subtasks, which have to operate on material and object characteristic/shape levels provided by accurately registered database. Moreover, the final fused and integrated application should be operated on objects of different sizes and scales, such as a single building detected within the urban area or a selected region on a building facade. The multi-scale and multi-sensor data fusion is possible with the eCognition procedure (user guide eCognition, 2003), when the substructures are archived by a hierarchical network.

The results of spectral/thermal classification processes are by far not only a spectral/thematic aggregation of classes converted to polygons or polylines (in vector format), but also a spatial and semantic structuring of the scene content (example of roofs extraction in Figure 11). The resulting network of extracted and identified objects can be seen as a spatial/semantic network of the scene. The local contextual information describes the joint relationships and meaningful interactions between those objects in the build urban environment and linked multi-scale and multi-sensor products. This hierarchy in the rule-base design allows a well-structured incorporation of knowledge.

Operation	Level	Object
Segmentation	Level 1	Roof

Hierarchical rule-base structure in eCognition:
Roof
- Shape 3-D
  - Shape
  - Proportions
    - Volume 3-D
    - Area 2-D
    - Diameter 2-D
  - Delta H (ground surface - top)

- Material
  - VNIR-SWIR thematic class
    - Tile
    - Bitumen
    - Concrete
  - Thermal micro – silicates class
    - Tile
    - Bitumen
    - Concrete

Fig. 11. Hierarchical rule-base structure in eCognition

In fact, now each object is identified not only by its spectral, thermal, textural, morphological, topological and shape properties, but also by its unique information linkage and its actual neighbors. The data is fused by mutual dependencies within and between objects that create a semantic network of the scene. To assure high level accuracy and operational efficiency the input products are inspected by the basic topological rule, which obligates that object borders overlay borders of objects on the next layer. Therefore, the multi-scale information, which is represented concurrently, can be related to each other.

The semantic network of fuzzy logic is an expert system that quantities uncertainties and variations of the input data. The fuzzy logic, as an alternative approach for the Boolean statements, avoids arbitrary thresholds and thus, it is able to estimate a real world environment (Benz et al., 2004). The implemented rules are guided by the reliability of class assignments, thus the solution is always possible, even if there are contradictory assignments (Civanlar & Trussel, 1986). This logic proposes a deliberate choice and parameterization of the membership function that established the relationship between object features and acceptable characteristics. Since the design is the most crucial step to introduce expert knowledge and information into the logic, the better and detailed the description of the real world environment are modeled by the membership function, the better the data fusion.

The operational system controls that a first class hierarchy will be loaded and used in the next step for data integration. Based on this preliminary fusion, first objects of interest are created from object primitives by thematic-based fusion. The same steps are performed until the final information (spectral quantitative model) is applied. The results are registered and integrated information is followed by the reliability map, which is established by the primary accuracy and classification confidence of each input data. The reliability map is important for post-processing inspection and testing routines; objects with low reliability must be assigned manually because no decision is possible. The suggested application involves semi-automatic or even manual stages, which have proven to be time-consuming operations. Yet, due to the expert system support, it is a time efficient application that produces highly accurate and reliable merged information.

5. Discussion

In this chapter, we present techniques for data fusion and data registration in several levels. Our study focused on the registration and the integration of multi-sensor and multi-temporal information for a 3-D urban environment monitoring application. For that purpose, both registration models and data fusion techniques were used.

The 3-D urban application satisfies a fundamental demand for non-traditional, multi-sensor and multi-type data. The frequent updating and extension requirement is replaced by integrating the variation in data formats and types for developing an urban environment. The main benefit of 3-D modeling and simulation over traditional 2-D mapping and analysis is a realistic illustration that can be regularly updated with attribute details and remote sensor-based quantitative/thermal information and models.

The proposed application offers an advanced methodology by integrating information into a 5-D data set. The ability to include an accurate and realistic 3-D position, quantitative information, thermal properties and temporal changes provide a near-real-time monitoring system for photogrammetric and urban planning purposes. The main objectives of many studies are linked to, and rely on a historical set of remotely sensed imagery for quantitative assessment and spatial evolution of an urban environment (Jensen and Cowen 1999, Donnay et al. 2001, Herold et al. 2003, 2005). The well-known methodology is pattern observation in the spatiotemporal and spectral domains. The main objective of this research is a fully controlled, near-real-time, natural and realistic monitoring system for an urban environment. This task led us first to combine the image-processing and map-matching procedures, and then incorporate remote sensing and GIS tools into an integrative method for data fusion and registration.

The proposed application for data fusion proved to be able to integrate several different types of data acquired from different sensors, and which are additionally dissimilar in rotation, translation, and possible scaling. The data fusion operated by fuzzy logic is a final product of the application. This approach is an important stage for quality assurance and validation but furthermore for information fusion in current and future remote sensing systems with multi-sensor sources.

The multi-dimensionality (5-D) of the developed urban environment application provides services such as thematic and thermal mapping, and a complete quantitative review of the building and its surroundings. These services are completed by providing the ability for accurate temporal monitoring and dynamic changes (changed detection) observations. The application design shows the possibility of delivering integrated information, and thus holistic views of whole urban environments, in a freeze-frame view of the spatio-temporal domain.

6. Conclusion

In conclusion, the suggested application may provide the urban planners, civil engineers and decision makers with tools to consider quantitative spectral information and temporal investigation in the 3-D urban space. It is seamlessly integrating the multi-sensor, multi-dimensional, multi-scaling and multi-temporal data into a 5-D operated system. The application provides a general overview of thematic maps, and the complete quantitative assessment for any building and its surroundings in a 3-D natural environment, as well as, the holistic view of urban environment.

7. Acknowledgment

This research work is supported by Discovery Grand (3-8163) from the Ministry of Science of Israel. The authors would like to express their deepest gratitude for this opportunity.

8. References

Akel, N.A.; Zilberstein, O. & Doytsher, Y. (2003). Automatic DTM extraction from dense raw LIDAR data in urban areas. In: Proc. FIG Working WeekParis, France, April 2003, 1-10.

Allen, J. & Lu, K. (2003). Modeling and prediction of future urban growth in the Charleston region of South Carolina: a GIS-based integrated approach, *Conservation Ecology*, 8(2), 202-211.

Ameri, B. (2000). Automatic recognition and 3-D reconstruction of buildings from digital imagery. Thesis (PhD), University of Stuttgart.

Benz, U.C.; Hofmann, P.; Willhauck, G.; Lingenfelder, I. & Heynen, M. (2004). Multi-resolution, object-oriented fuzzy analysis of remote sensing data for GIS-ready information. *ISPRS Journal of Photogrammetry & Remote Sensing*, 58, 239– 258

Brook, A. & Ben-Dor, E. (2011[a]). Advantages of boresight effect in the hyperspectral data analysis. *Remote Sensing*, 3 (3), 484-502.

Brook, A. & Ben-Dor, E. (2011[b]). Supervised vicarious calibration of hyperspectral remote sensing data. *Remote Sensing of Environment*, 115, 1543-1555.

Brook, A. & Ben-Dor, E. (2011[c]). Automatic registration of airborne and space-borne images by topology map-matching with SURF. *Remote Sensing*, 3, 65-82.

Brook, A.; Ben-Dor, E. & Richter, R. (2011). Modeling and monitoring urban built environment via multi-source integrated and fused remote sensing data. *International Journal of Image and Data Fusion*, in press, 1-31.

Brown, L.G. (1992). A survey of image registration techniques. *ACM Computing Surveys*, 24, 325-376.

Brown, H. & Lowe, D. (2002). Invariant features from interest point groups, in BMVC.

Bulmer, D. (2001). How can computer simulated visualizations of the built environment facilitate better public participation in the planning process? *On Line Planning Journal*, 1-28, http://www.onlineplaning.org

Carlson, T.N.; Dodd, J.K.; Benjamin, S.G. & Cooper, J.N. (1981). Satellite estimation of the surface energy balance, moisture availability and thermal inertia. *Journal of Applied Meteorology*, 20, 67–87.

Chan, R.; Jepson, W. & Friedman, S. (1998). Urban simulation: an innovative tool for interactive planning and consensus building. In: Proceedings of the 1998 American Planning Association National Conference, Boston, MA.

Chantous, M.; Ghosh, S. & Bayoumi, M.A. (2009). Multi-modal automatic image registration technique based on complex wavelets. In: Proceedings of the 16th IEEE International Conference on Image Processing, Cairo, Egypt, 173-176.

Civanlar, R. & Trussel, H. (1986). Constructing membership functions using statistical data. *IEEE Fuzzy Sets and Systems*,18, 1 –14.

Cloude, S.P.; Kootsookos, P.J. & Rottensteiner, F. (2004). The automatic extraction of roads from LIDAR data. In: ISPRS 2004, Istanbul, Turkey.

Conel, J.E. (1969). Infrared Emissivities of Silicates: Experimental Results and a Cloudy Atmosphere Model of Spectral Emission from Condensed Particulate Mediums. Journal of Geophysical Research, 74 (6), 1614-1634.

Cressie, A.N.C. (1993). Statistics for spatial data. Review. New York: Wiley.

Dodge, M.; Smith, A. & Fleetwood, S., 1998. Towards the virtual city: VR & internet GIS for urban planning. In: Virtual Reality and Geographical Information Systems. London: Birkbeck College.

Donnay, J.P.; Barnsley, M.J. & Longley, P.A. (2001). Remote sensing and urban analysis. In: J.P. Donnay, M.J. Barnsley and P.A. Longley, eds. *Remote sensing and urban analysis*. London and New York: Taylor and Francis, 3-18.

Fan, X., Rhody, H. and Saber, E., 2005. Automatic registration of multi-sensor airborne imagery. In: Proceedings of the 34th Applied Imagery and Pattern Recognition Workshop, Washington DC, 80-86.

Heiden, U., Segl, K., Roessner, S. and Kaufmann, H., 2007. Determination of robust spectral features for identification of urban surface materials in hyperspectral remote sensing data. *Remote Sensing of Environment*, 111, 537-552.

Henry, J.A.; Dicks, S.E.; Wetterqvist, O.F. & Roguski, S.J. (1989). Comparison of satellite, ground-based, and modeling techniques for analyzing the urban heat island. *Photogrammetric Engineering and Remote Sensing*, 55, 69–76.

Herold, M., Goldstein, N.C. and Clarke, K.C., 2003. The spatiotemporal form of urban growth: measurement, analysis and modeling. *Remote Sensing of Environment*, 86, 286-302.

Herold, M., Couclelis, H. and Clarke, K.C., 2005. The role of spatial metrics in the analysis and modeling of land use change. Computers, *Environment and Urban Systems*, 29(4), 369-399.

Hinz, S. and Baumgartner, A., 2003. Automatic extraction of urban road networks from multi-view aerial imagery. *ISPRS Journal of Photogrammetry and Remote Sensing*, 58 (1-2), 83-98.

Jarvis, R.A., 1973. On the identification of the convex hull of a finite set of points in the plane. *Information Processing Letters*, 2, 18-21.

Jarvis, R.A., 1977. Computing the shape hull of points in the plane. In: Proceedings of the IEEE Computer Society Conference Pattern Recognition and Image Processing, 231-241.

Jensen, J.R. and Cowen, D.C., 1999. Remote sensing of urban/suburban infrastructure and socio-economic attributes. *Photogrammetric Engineering and Remote Sensing*, 65 (5), 611-622.

Jensen, J.R., 2004. Introductory digital image processing. 3rd ed. Upper Saddle River, NJ: Prentice Hall.

Jepson, W.H., Liggett, R.S. and Friedman, S., 2001. An integrated environment for urban simulation. In: R.K. Brail and R.E. Klosterman, eds. Planning support systems: integrating geographic information systems, models, and visualization tools. Redlands, CA: ESRI, 387-404.

Juan, G., Martinez, M. and Velasco, R., 2007. Hyperspectral remote sensing application for semi-urban areas monitoring. *Urban Remote Sensing Joint Event*, 11 (13), 1-5.

Kidder, S.Q. & Wu, H-T. (1987). A multispectral study of the St. Louis area under snow-covered conditions using NOAA-7 AVHRR data. *Remote Sensing of Environment*, 22, 159-172.

Kolbe, T.H., Gerhard, G. and Plümer, L., 2005. CityGML—Interoperable access to 3D city models. In: International Symposium on Geoinformation for Disaster Management GI4DM 2005, Delft, Netherlands, Lecture Notes in Computer Science, March, 2005.

Kolbe, T. and Bacharach, S., 2006. CityGML: An open standard for 3D city models. *Directions Magazine ESRI*, http://directionmag.com/articles/123103

Lee, H.Y., Park, W., Lee, H.-K. and Kim, T.-G., 2000. Towards knowledge-based extraction of roads from 1m resolution satellite images. In: Proceedings of the IEEE Southwest Symposium on Image Analysis and Interpretation, Austin, TX, 171-176.

Li, R. and Zhou, G., 1999. Experimental study on ground point determination from high-resolution airborne and satellite imagery. In: Proceedings of the ASPRS Annual Conference, Portland, ME, 88-97.

Li, Y., 2008. Automated georeferencing. Thesis (PhD). University of Texas at Dallas.

Lindeberg, T., 2004. Feature detection with automatic scale selection. *International Journal of Computer Vision*, 30, 79-116.

Masaharu, H. and Ohtsubo, K., 2002. A filtering method of airborne laser scanner data for complex terrain. *The International Archives of Photogrammetry, Remote Sensing, and Spatial Information Sciences*, 15 (3B), 165-169.

Moigne, J.L., Campbel, W.J. and Cromp, R.F., 2002. An automated parallel image registration technique based on the correlation of wavelet features. *IEEE Transactions on Geoscience and Remote Sensing*, 40, 1849-1864.

Nichol, JE. (1994). A GIS-based approach to microclimate monitoring in Singapore's high-rise housing estates. *Photogrammetric Engineering and Remote Sensing*, 60, 1225-1232.

Nichol, J.E. (1996). High-resolution surface temperature patterns related to urban morphology in a tropical city: a satellite-based study. *Journal of Applied Meteorology*, 35, 135-146.

Pauca,V.P.; Piper, J. & Plemmons R.J. (2006) Nonnegative matrix factorization for spectral data analysis. *Linear Algebra and Applications*, 416(1), 29-47.

Pudil, P.; Novovicova, J. & Kittler, J. (1994). Floating search methods in feature selection, *Pattern Recognition Letters*, 15, 1119 – 1125.

Ridd, M.K. 1995. Exploring V-I-S model for urban ecosystem analysis through remote sensing. *International Journal of Remote Sensing*, 16, 993-1000.

Richards, J.A. and Jia, X., 1999. Remote sensing digital image analysis: an introduction. New York: Springer-Verlag.

Robila, S.A. & Maciak, L.G. (2006). Considerations on Parallelizing Nonnegative Matrix Factorization for Hyperspectral Data Unmixing, *IEEE Geoscience and Remote Sensing Letters*, 6(1), 57 – 61.

Roessner, S., Segl, K., Heiden, U., Munier, K. and Kaufmann, H., 1998. Application of hyperspectral DAIS data for differentiation of urban surface in the city of Dresden, Germany. In: Proceedings 1st EARSel Workshop on Imaging Spectroscopy, Zurich, 463-472.

Roessner, S., Segl, K., Heiden, U. and Kaufmann, H., 2001. Automated differentiation of urban surfaces based on airborne hyperspectral imagery. *IEEE Transactions on Geoscience and Remote Sensing*, 39 (7), 1525-1532.

Roth, M.; Oke, T.R. & Emery, W.J. (1989). Satellite-derived urban heat islands from three coastal cities and the utilization of such data in urban climatology. *International Journal of Remote Sensing*, 10, 1699–1720.

Rottensteiner, F., Trinder, J., Clode, S., Kubic, K., 2003. Building detection using LIDAR data and multispectral images. In: Proceedings of DICTA, Sydney, Australia, 673-682.

Sacks, J.; Welch, W.J.; Mitchell, T.J. &Wynn, H.P. (1989). Design and analysis of computer experiments. *Statistical Science*, 4(4), 409–435.

Shan, J. and Sampath, A., 2005. Urban DEM generation from raw LIDAR data: a labeling algorithm and its performance. *Photogrammetric Engineering and Remote Sensing*, 71 (2), 217-226.

Song, Y., Wang, H., Hamilton, A. and Arayici, Y., 2009. Producing 3D applications for urban planning by integrating 3D scanned building data with geo-spatial data. Protocol. Research Institute for the Built and Human Environment (BuHu), University of Salford, UK.

Tao, V., 2001. Database-guided automatic inspection of vertically structured transportation objects from mobile mapping image sequences. *In: ISPRS Press*, 1401-1409.

UserGuide eCognition, 2003. Website: www.definiens_imaging.com.

Velaga, N.R., Quddus, M.A. and Bristow, A.L., 2009. Developing an enhanced weight-based topological map-matching algorithm for intelligent transport systems. *Transportation Research Part C: Emerging Technologies*, 17, 672-683.

Villa, A.; Chanussot, J.; Benediktsson, J.A. & Jutten, C. (2011). Spectral Unmixing for the Classification of Hyperspectral Images at a Finer Spatial Resolution. *IEEE Selected Topics in Signal Processing*, 5(3), 521 – 533.

Viola, P. and Wells, W.M., 1997. Alignment by maximization of mutual information. *International Journal of Computer Vision*, 24, 137-154.

Vukovich, F.M. (1983). An analysis of the ground temperature and reflectivity pattern about St. Louis, Missouri, using HCMM satellite data. *Journal of Climate and Applied Meteorology*, 22, 560–571.

Wang, Y., 2008. A further discussion of 3D building reconstruction and roof reconstruction based on airborne LiDAR data by VEPS' partner, the Department of Remote Sensing and Land Information Systems, Freiburg.

Wang, J. and Shan, J., 2009. Segmentation of LiDAR point clouds for building extraction. In: ASPRS 2009 Annual Conference, Baltimore, MD.

Whitney, A.W. (1971). A Direct Method of Nonparametric Measurement Selection, *IEEE Trans. Computers*, 20(9), 1100-1103.

Wu, J. and Chung, A., 2004. Multimodal brain image registration based on wavelet transform using SAD and MI. In: Proceedings of the 2nd International Workshop on Medical Imaging and Augmented Reality, Beijing, China.

Xu, R. and Chen, Y., 2007. Wavelet-based multiresolution medical image registration strategy combining mutual information with spatial information. International Journal of Innovative Computing, Information and Control, 3, 285-296.

Yang, F. & Jiang, T. (2003). Pixon-Based Image Segmentation With Markov Random Fields. *IEEE Transactions on Image Processing*, 12, 1552-1559.

Young, S.J., Johnson, R.B., and Hackwell, J.A., 2002. An in-scene method for atmospheric compensation of thermal hyperspectral data. *Journal of Geophysical Research*, 107, 20-28.

Zhang K, Chen S, Whitman D, Shyu M, Yan J, Zhang C. 2003. A Progressive Morphological Filter for Removing Non-Ground Measurements from Airborne LIDAR Data. *IEEE Transactions on Geoscience and Remote Sensing*, 41(4), 872-882.

Zavorin, I. and Le Moigne, J., 2005. Use of multiresolution wavelet feature pyramids for automatic registration of multisensor imagery. *IEEE Transactions on Image Processing*, 14, 770-782.

Zhou, G., 2004. Urban 3D GIS from LiDAR and digital aerial images. *Computers and Geosciences*, 30, 345-353.

Zitova, B. and Flusser, J., 2003. Image registration methods: a survey. *Image and Vision Computing*, 21, 977-1000.

5

Estimation of the Separable MGMRF Parameters for Thematic Classification

Rolando D. Navarro, Jr., Joselito C. Magadia and Enrico C. Paringit
University of the Philippines, Diliman, Quezon City
Philippines

1. Introduction

Because of its ability to describe interdependence between neighboring sites, the Markov Random Field (MRF) is a very attractive model in characterizing correlated observations (Moura and Balram, 1993) and it has potential applications in areas of remote sensing, such as spatio-temporal modeling and machine vision. In this study, we model image random field conditional to the texture label as a Multivariate Gauss Markov Random Field (MGMRF); whereas; the thematic map is modeled as a discrete label MRF (Li, 1995). The observations in the Gauss Markov Random Field (GMRF) are distributed with the Gaussian distribution.

There are some MGMRF models where the interaction matrices are modeled in some simplified form, including the MGMRF with isotropic interaction matrix which we shall refer here as Hazel's GMRF (Hazel, 2000), The MGFMRF with anisotropic interaction matrix proportional to the identity matrix which we shall refer here as Rellier's GMRF (Rellier et. al., 2004), and the Gaussian Symmetric Clustering (GSC) (Hazel, 2000).

From these developments, the model for anisotropic GMRF was generalized and its parameter estimator for an arbitrary neighborhood system is characterized (Navarro et al., 2009). Using our model, the classification performance was analyzed and compared with the GMRF models in literature.

Spectral classes are explored in segmenting image random field models to be able to extract the spatial, spectral, and temporal information. A special case is addressed when the observation includes spectral and temporal information known as the spectro-temporal observation. With respect to the spectral and temporal dimensions, the separability structure is considered based on the Kronecker tensor product of the GMRF model parameters. Separable parameters contain less parameters, compared with its non-separable counterpart. In addition, the spectral and temporal dimensions on a separable model can be analyzed separately. We analyzed whether the separability of the GMRF parameters would improve the classification of the thematic map.

2. Image random field modelling and thematic classification

This section covers statistical background in characterizing random fields based on the MRF. Then, we will present estimation for the thematic map and image random field parameters.

Finally the thematic map classifier is presented based on the Iterated Conditional Modes (ICM) algorithm.

2.1 Markov random fields

A random field $\mathbf{Z} = \left\{\mathbf{Z_s} : s \in \mathcal{S}\right\}$ where s is a site on the lattice $\mathcal{S}$ with the neighborhood system ∂ with parameter Π is a MRF if for $s \in \mathcal{S}$ (Winkler, 2003).

$$p\left(\mathbf{Z_s}\middle|\mathbf{Z}_{\mathcal{S}/s};\Pi\right) = p\left(\mathbf{Z_s}\middle|\mathbf{Z}_{\partial s};\Pi\right) \tag{1}$$

where $\mathbf{Z}_{\partial s} = \left\{\mathbf{Z_t} : t \in \partial s\right\}$ is the random field which consists of observations of the neighbors of s. Similarly, $\mathbf{Z}_{\mathcal{S}/s} = \left\{\mathbf{Z_t} : t \in \mathcal{S}/s\right\}$ is the random field, which consists of observations that exclude s.

2.2 Thematic map modeling

Let $L = \left\{L_s\right\}_{s \in \mathcal{S}}$ be denoted as the thematic map, where $L_s \in \left\{1, \cdots, M\right\}$ is the labeled thematic class at site s and M is the number of thematic classes. The thematic map is modeled as a discrete space, discrete domain MRF with parameters $\varphi = \left\{\left\{a_m\right\}_{1 \leq m \leq M}, \left\{b_r\right\}_{r \in \mathcal{N}}\right\}$ where a_m is the singleton potential coefficient for the mth thematic class, b_r are made up by the pairwise potential coefficients, and $\mathcal{N}$ is region of support (Jeng & Woods, 1991) or the neighborhood set (Kasyap & Chellappa, 1983). Its conditional probability density function (pdf) is given by

$$p\left(L_s\middle|L_{\partial s},\varphi\right) = \frac{\exp\left(\sum\limits_{m=1}^{M} a_m \mathbf{1}_{\{\mathbf{L_s}=m\}} + \sum\limits_{r \in \mathcal{N}} b_r \cdot V\left(L_s, L_{s-r}\right)\right)}{\sum\limits_{l=1}^{M} \exp\left(a_l + \sum\limits_{r \in \mathcal{N}} b_r \cdot V\left(L_s = l, L_{s-r}\right)\right)} \tag{2}$$

(Li, 1995), where

$$V\left(x,y\right) = \begin{cases} 1 & x = y \\ -1 & x \neq y. \end{cases}$$

2.3 Image random field modeling

The observation $\mathbf{Y_s}$ given the thematic map $\mathbf{L}$ is modeled with the conditional distribution $\mathbf{Y_s}\middle|\mathbf{L} \sim N_N\left(\mu\left(L_s\right), \Sigma\left(L_s\right)\right)$. It is conditionally dependent on L_s, the thematic class at site s, and it is driven by an autoregressive Gaussian colored noise process $\mathbf{X_s}\middle|\mathbf{L} \sim N_N\left(0_{N \times 1}, \Sigma\left(L_s\right)\right)$. Two noise processes $\mathbf{X_s}$ and $\mathbf{X_{s-r}}$ are statistically independent if the corresponding thematic classes L_s and L_{s-r} are different for all $r \in \mathcal{N}$ and $s \in \mathcal{S}$. This model tends to avoid the blurring effect created between segment boundaries which, in turn, may yield poor classification performance. The resulting equation can be written as follows:

$$\mathbf{X_s} = \left(\mathbf{Y_s} - \mu\left(L_s\right)\right) - \sum\limits_{r \in \mathcal{N}} \theta_r\left(L_s\right)\mathbf{1}_{\{L_s = L_{s-r}\}}\left(\mathbf{Y_{s-r}} - \mu\left(L_s\right)\right). \tag{3}$$

The noise process has the following characterization:

$$E\left[\mathbf{X_s}|\mathbf{L};\mathbf{\Theta}\right] = \mathbf{0}_{N\times 1} \tag{4}$$

$$\mathrm{cov}\left(\mathbf{X_s},\mathbf{X_{s-r}}|\mathbf{L};\mathbf{\Theta}\right) = \begin{cases} \mathbf{\Sigma}(L_s) & \mathbf{r} = \mathbf{0}_{p\times 1} \\ -\mathbf{\theta_r}(L_s)\mathbf{\Sigma}(L_s)\mathbf{1}_{\{L_s=L_{s-r}\}} & \mathbf{r}\in\mathcal{N} \\ \mathbf{0}_{N\times N} & \text{otherwise} \end{cases} \tag{5}$$

$$\mathrm{cov}\left(\mathbf{X_s},\mathbf{Y_{s-r}}|\mathbf{L};\mathbf{\Theta}\right) = \mathbf{\Sigma}(L_s)\cdot\mathbf{1}_{\{\mathbf{r}=\mathbf{0}_{p\times 1}\}}. \tag{6}$$

The conditional probability on the other hand is given as

$$p\left(\mathbf{Y_s}|\mathbf{Y}_{\partial s},\mathbf{L};\mathbf{\Theta}\right) = \frac{1}{(2\pi)^{N/2}|\mathbf{\Sigma}(L_s)|^{1/2}}\exp\left(-\frac{1}{2}\mathbf{X_s}^T\mathbf{\Sigma}^{-1}(L_s)\mathbf{X_s}\right). \tag{7}$$

2.4 Maximum pseudo-likelihood estimation

The maximum pseudo-likelihood estimation (MPLE) combines sites to form the pseudo-likelihood function from the conditional probabilities (Li, 1995). The pseudo-likelihood functions for the thematic map random field and image random field parameters are given as follows:

$$PL(\varphi) = \prod_{s\in\mathcal{S}} p\left(L_s|\mathbf{L}_{\partial s};\varphi\right) \tag{8}$$

$$PL(\mathbf{\Theta}|\mathbf{L}) = \prod_{m=1}^{M}\prod_{s\in\mathcal{S}(m)} p\left(\mathbf{Y_s}|\mathbf{Y}_{\partial s},\mathbf{L};\mathbf{\Theta}\right) \tag{9}$$

where $\mathcal{S}(m)$ is the collection of sites with the mth thematic class. The MPLE possesses an invariance property, that is, if $\hat{\mathbf{\Pi}}$ is the MPLE of the parameter $\mathbf{\Pi}$, then for an arbitrary function τ, $\tau(\hat{\mathbf{\Pi}})$ is the MPLE of the parameter $\tau(\mathbf{\Pi})$. The proof is similar to that of the invariance property of the MLE (Casella and Berger, 2002) since the form of the pseudo-likelihood function is analogous that of the likelihood function, depending on the parameter given the data. Moreover, the MPLE converges to the MLE almost surely as the lattice size approaches infinity (Geman and Greffigne, 1987).

2.5 Thematic classification

The thematic map can be recovered by the maximum a posteriori probability (MAP) rule. It can be implemented using a numerical optimization technique such as Simulated Annealing (SA) (Jeng & Woods, 1991). Although the global convergence employing SA is guaranteed almost surely, its convergence is very slow (Aarts & Korts, 1987; Winkler, 2006). An alternative to this is to use the ICM algorithm (Besag, 1986) given as

$$\hat{L}_\mathbf{s} = \arg\max_{1\leq m\leq M} p\left(\mathbf{Y}|L_\mathbf{s}=m,\mathbf{L}_{\mathcal{S}/\mathbf{s}};\mathbf{\Theta}\right)p\left(L_\mathbf{s}=m|\mathbf{L}_{\mathcal{S}/\mathbf{s}};\varphi\right). \tag{10}$$

This is interpreted as the instantaneous freezing of the annealing schedule of the SA. However, since $p(\mathbf{Y}|\mathbf{L};\Theta)$ is difficult to evaluate, alternatively, it is replaced by its pseudo-likelihood (Hazel, 2000) given as

$$p(\mathbf{Y}|\mathbf{L};\Theta) \approx \prod_{s \in \mathcal{S}} p(\mathbf{Y_s}|\mathbf{Y}_{\partial s}, \mathbf{L};\Theta). \tag{11}$$

Hence, the classifier is reduced to

$$\hat{L}_\mathbf{s} = \arg \max_{1 \le m \le M} \prod_{s \in \mathcal{S}} p(\mathbf{Y_s}|\mathbf{Y}_{\partial s}, \mathbf{L};\Theta) \cdot p(L_\mathbf{s} = m|\mathbf{L}_{\mathcal{S}/s};\boldsymbol{\varphi}). \tag{12}$$

The ICM algorithm, unlike the SA, is only guaranteed to converge to the local maxima. This problem can be alleviated by initializing the thematic map from the Gaussian Spectral Clustering (GSC) model (Hazel, 2000).

2.6 Numerical implementation

The MPLE-based estimators are not in their closed form and must be evaluated numerically. The pseudocode for estimating the parameters is presented below.

Initialize $\mathbf{L}$, $\boldsymbol{\varphi}$, and Θ
Estimate $\boldsymbol{\varphi}$
Estimate Θ
    Estimate $\boldsymbol{\mu}(m)$ given $\boldsymbol{\theta}_r(m)$ and $\boldsymbol{\Sigma}(m)$
    Estimate $\boldsymbol{\theta}_r(m)$ given $\boldsymbol{\Sigma}(m)$ and $\boldsymbol{\mu}(m)$
    Estimate $\boldsymbol{\Sigma}(m)$ given $\boldsymbol{\mu}(m)$ and $\boldsymbol{\theta}_r(m)$
Estimate L_s by the ICM Algorithm

The image random field parameters are estimated using a method with some resemblance to the Gauss-Seidel iteration method (Kreyzig, 1993). The convergence criterion for estimating these parameters using this iteration method has yet to be established. As a precautionary measure, a single iteration was performed. This method was also applied in estimating the image random field estimators in Rellier's GMRF (Rellier, et. al., 2004).

3. Spectro-temporal MGMRF modelling

The spectro-temporal observation image random field will be characterized with hybrid separable MGMRF parameters.

3.1 Image random field modeling

We let M_1 - number of lines, M_2 - number of samples, N_1 - number of spectral bands, and N_2 - number of temporal slots. The image random field is characterized as follows:

Lattice $\mathcal{S} = \{(s_1, s_2): 1 \le s_1 \le M_1, 1 \le s_2 \le M_2\}$

Thematic Class $L_\mathbf{s} = L_{(s_1, s_2)}$

The thematic class L_s at a given site $\mathbf{s} \in \mathcal{S}$ is modeled to be fixed over time.

Observation
$$\mathbf{Y}_s = \mathbf{Y}_{(s_1,s_2)} = \left(W_{(s_1,s_2,1,1)} \quad \cdots \quad W_{(s_1,s_2,k,l)} \quad \cdots \quad W_{(s_1,s_2,N_1,N_2)} \right)^T$$

The observation $\mathbf{Y}_s$ is a multispectral and mono-temporal vector of reflectance of the given spatial location (s_1, s_2) measured at the kth spectral band with wavelength $\lambda = \lambda_k$, for $1 \le k \le N_1$, and at the lth temporal slot with time $T = T_l$ for $1 \le l \le N_2$. More specifically, the $\left(k + (l-1)N_1 \right)^{th}$ element of $\mathbf{Y}_s$ denoted as $Y_{s,(k,l)}$ is given as $Y_{s,(k,l)} = W_{(s_1,s_2,k,l)}$.

Let us consider the matrix $\mathbf{Y}_s^{\#}$ defined by rearranging the elements of the spectro-temporal observation $\mathbf{Y}_s$ with the reshape operator $\mathbf{Y}_s^{\#} = reshape\left(\mathbf{Y}_s, N_1, N_2 \right)$. The reshape function given as $\mathbf{B} = reshape\left(\mathbf{A}, N_1, N_2 \right)$ transforms the vector $\mathbf{A} = \{a_k\} \in \mathbb{R}^{N_1 N_2}$ into the $N_1 \times N_2$ matrix $\mathbf{B} = \{b_{ij}\} \in \mathbb{R}^{N_1 \times N_2}$ by the mapping $b_{ij} = a_{k=i+(j-1)N_1}$ for all $1 \le i \le N_1$ and $1 \le j \le N_2$, i.e.

$$\mathbf{Y}_s^{\#} = \begin{bmatrix} Y_{s,(1,1)} & Y_{s,(1,2)} & \cdots & Y_{s,(1,N_2)} \\ Y_{s,(2,1)} & Y_{s,(2,2)} & \cdots & Y_{s,(2,N_2)} \\ \cdots & \cdots & \cdots & \cdots \\ Y_{s,(N_1,1)} & Y_{s,(N_1,2)} & \cdots & Y_{s,(N_1,N_2)} \end{bmatrix}. \tag{13}$$

The matrix $\mathbf{Y}_s^{\#}$ is characterized by allocating the reflectance across the bands for a given time by column and the reflectance across time for a given band by row.

3.2 Separable structure of the covariance matrix

There is a growing interest in modeling the covariance structure with more than one attribute. For example, in spatio-temporal modeling, the covariance structure of "spatial" and "temporal" attributes is jointly considered (Kyriakidis and Journel, 1999; Huizenga, et. al., 2002). On the other hand, in the area of longitudinal studies the covariance structure between "factors" and "temporal" attributes are jointly considered (Naik and Rao, 2001). Both studies mentioned above considered covariance matrices with a separable structure between these attributes.

In the realm of remote sensing, few studies have been conducted combining the covariance structure involving spectro-temporal attributes. Campbell and Kiiveri demonstrated canonical variates calculations are reduced to simultaneous between-groups and within-group analyses of a linear combination of spectral bands over time, and the analyses of a linear combination of the time over the spectral bands (Campbell and Kiiveri, 1988).

In light of recent literature, we propose to model the GMRF models as applied to remote sensing image processing where the covariance structure of the "spectral" and "temporal" attributes is characterized jointly. The separable covariance structure associated with the matrix Gaussian distribution has been considered.

3.2.1 Non-separable covariance structure

The matrix observation driven by a colored noise and its vectorized distribution, is assumed to be a realization from the process whose conditional form is given by

$\mathbf{X}_s|\mathbf{L} \sim N_N\left(\mathbf{0}_N, \Sigma(L_s)\right)$. The covariance matrix $\Sigma(L_s)$ does not have any special structure, except it has to be a positive definite symmetric matrix. This covariance matrix structure referred to as an unpatterned covariance matrix (Dutilleul, 1999). The statistical characterization is similar to the MGMRF discussed in Section 2.3.

3.2.2 Matrix gaussian distribution

Let $\mathbf{X}^\#$ be a random matrix distributed as $\mathbf{X}^\# \sim N_{m;n}\left(\mathbf{M}^\#, \Xi^{(1)}, \Xi^{(2)}\right)$ where $\mathbf{M}^\# \in \mathbb{R}^{m\times n}$ is the expectation matrix, $\Xi^{(1)} \in \mathbb{R}^{m\times m}$ is the covariance matrix across the rows, and $\Xi^{(2)} \in \mathbb{R}^{n\times n}$ is the covariance matrix across the columns. Hence, the pdf of $\mathbf{X}^\#$ is given as

$$p\left(\mathbf{X}^\#\right) = \frac{1}{(2\pi)^{mn/2}\left|\Xi^{(1)}\right|^{n/2}\left|\Xi^{(2)}\right|^{m/2}} \exp\left[-\frac{1}{2}tr\left(\left(\Xi^{(1)}\right)^{-1}\left(\mathbf{X}^\# - \mathbf{M}^\#\right)\left(\Xi^{(2)}\right)^{-1}\left(\mathbf{X}^\# - \mathbf{M}^\#\right)^T\right)\right] \quad (14)$$

(Arnold, 1981). Also, if we stack the matrix $\mathbf{X}^\#$ into the random vector $\mathbf{X} \equiv vec\left(\mathbf{X}^\#\right)$, then $\mathbf{X} \sim N_{mn}\left(\mathbf{M}, \Xi\right)$ where $\mathbf{M} = vec\left(\mathbf{M}^\#\right) \in \mathbb{R}^{mn}$ is the expectation matrix and $\Xi = \Xi^{(2)} \otimes \Xi^{(1)} \in \mathbb{R}^{mn\times mn}$ is the covariance matrix (Arnold, 1981), and its pdf is given as

$$p(\mathbf{X}) = \frac{1}{(2\pi)^{mn/2}|\Xi|^{1/2}} \exp\left[-\frac{1}{2}(\mathbf{X}-\mathbf{M})^T \Xi^{-1}(\mathbf{X}-\mathbf{M})\right]. \quad (15)$$

We model the associated noise process $\mathbf{X}_s^\#$ as a matrix Gaussian distribution, i.e. $\mathbf{X}_s^\#|\mathbf{L} \sim N_{N_1,N_2}\left(\mathbf{0}_{N_1\times N_2}, \Sigma^{(1)}(L_s), \Sigma^{(2)}(L_s)\right)$ where $\Sigma^{(1)}(L_s) \in \mathbb{R}^{N_1\times N_1}$ is the covariance matrix across the bands, and $\Sigma^{(2)}(L_s) \in \mathbb{R}^{N_2\times N_2}$ is the covariance matrix across time. Stacking the matrix $\mathbf{X}_s^\#$ into a random vector $\mathbf{X}_s \equiv vec\left(\mathbf{X}_s^\#\right) \in \mathbb{R}^{N_1N_2}$ corresponds to the vectorized colored noise with conditional distribution $\mathbf{X}_s|\mathbf{L} \sim N_{N_1N_2}\left(\mathbf{0}_{N_1N_2}, \Sigma^{(2)}(L_s) \otimes \Sigma^{(1)}(L_s)\right)$.

3.2.3 Separable covariance structure

The spectro-temporal, separable covariance matrix model (Lu and Zimmerman, 2005; Fuentes, 2006) has the form

$$\Sigma(m) = \Sigma^{(2)}(m) \otimes \Sigma^{(1)}(m) \quad (16)$$

for $1 \leq m \leq M$ where $\Sigma^{(1)}(m) = \left\{\sigma_{ij}^{(1)}(m)\right\} \in \mathbb{R}^{N_1\times N_1}$ is the covariance matrix across bands and $\Sigma^{(2)}(m) = \left\{\sigma_{kl}^{(2)}(m)\right\} \in \mathbb{R}^{N_2\times N_2}$ is the covariance matrix across time. Now, since

$$\mathbf{X}_s|\mathbf{L} \sim N_{N_1N_2}\left(\mathbf{0}_{N_1\times N_2}, \Sigma^{(2)}(L_s) \otimes \Sigma^{(1)}(L_s)\right) \quad (17)$$

$$\mathbf{Y}_s|\mathbf{L} \sim N_{N_1N_2}\left(\boldsymbol{\mu}(L_s), \Sigma^{(2)}(L_s) \otimes \Sigma^{(1)}(L_s)\right) \quad (18)$$

then, the covariance is given as (Arnold, 1981):

$$\mathrm{cov}\left(X_{s,(k,l)}, X_{s,(k,l)} \middle| \mathbf{L}; \Theta\right) = \sigma_{kk}^{(1)}(L_s)\sigma_{ll}^{(2)}(L_s) \tag{19}$$

$$\mathrm{cov}\left(Y_{s,(k,l)}, Y_{s,(k,l)} \middle| \mathbf{L}; \Theta\right) = \sigma_{kk}^{(1)}(L_s)\sigma_{ll}^{(2)}(L_s) . \tag{20}$$

This corresponds to the product of the variance associated with the reflectance at the kth spectral band $\sigma_{kk}^{(1)}(L_s)$ and the variance associated with the reflectance at the lth temporal slot $\sigma_{ll}^{(2)}(L_s)$. Likewise, the cross-covariance is given as (Arnold, 1981):

$$\mathrm{cov}\left(X_{s,(i,j)}, X_{s,(k,l)} \middle| \mathbf{L}; \Theta\right) = \sigma_{ik}^{(1)}(L_s)\sigma_{jl}^{(2)}(L_s) \tag{21}$$

$$\mathrm{cov}\left(Y_{s,(i,j)}, Y_{s,(k,l)} \middle| \mathbf{L}; \Theta\right) = \sigma_{ik}^{(1)}(L_s)\sigma_{jl}^{(2)}(L_s) . \tag{22}$$

This corresponds to the product of the covariance associated with the reflectance at the ith and the kth spectral band $\sigma_{ik}^{(1)}(L_s)$ and the covariance associated with the reflectance at the jth and the lth temporal slot $\sigma_{jl}^{(2)}(L_s)$.

The number of parameters in the unpatterned covariance matrix is $N(N+1)/2 = N_1 N_2 (N_1 N_2 + 1)/2$. On the other hand, the number of parameters for a separable covariance matrix is $\left[N_1(N_1+1) + N_2(N_2+1)\right]/2$, which has fewer parameters compared to its non-separable counterpart.

3.2.4 Separable of interaction matrix structure

We can also model the interaction matrix coefficients with a separable structure for all $\mathbf{r} \in \mathcal{N}$ and $1 \le m \le M$ of the form

$$\theta_{\mathbf{r}}(m) = \theta_{\mathbf{r}}^{(2)}(m) \otimes \theta_{\mathbf{r}}^{(1)}(m) \tag{23}$$

where $\theta_{\mathbf{r}}^{(1)}(m) \in \mathbb{R}^{N_1 \times N_1}$ is the interaction matrix across the bands and $\theta_{\mathbf{r}}^{(2)}(m) \in \mathbb{R}^{N_2 \times N_2}$ is the interaction matrix across time. In the next section, the interaction matrix coefficient $\theta_{\mathbf{r}}(m)$ can be made separable for $\mathbf{r} \in \mathcal{N}$ and $1 \le m \le M$ provided that $\Sigma(m)$ is separable. Furthermore, if $\Sigma(m)$ is separable, then the following is the resulting statistical characterization of $\mathbf{X}_s$:

$$E\left[\mathbf{X}_s \middle| \mathbf{L}; \Theta\right] = \mathbf{0}_{N \times 1} \tag{24}$$

$$\mathrm{cov}\left(\mathbf{X}_s, \mathbf{X}_{s-\mathbf{r}} \middle| \mathbf{L}; \Theta\right) = \begin{cases} \Sigma^{(2)}(L_s) \otimes \Sigma^{(1)}(L_s) & \mathbf{r} = \mathbf{0}_{p \times 1} \\ -\left(-\theta_{\mathbf{r}}^{(2)}(L_s)\Sigma^{(2)}(L_s) \cdot \mathbf{1}_{\{L_s = L_{s-\mathbf{r}}\}}\right) \otimes \left(-\theta_{\mathbf{r}}^{(1)}(L_s)\Sigma^{(1)}(L_s) \cdot \mathbf{1}_{\{L_s = L_{s-\mathbf{r}}\}}\right) & \mathbf{r} \in \mathcal{N} \\ \mathbf{0}_{N_2} \otimes \mathbf{0}_{N_1} & \text{otherwise} \end{cases} \tag{25}$$

$$\text{cov}\left(\mathbf{X_s},\mathbf{Y_{s-r}}|\mathbf{L};\Theta\right)=\Sigma^{(2)}\left(L_s\right)\cdot\mathbf{1}_{\{L_s=L_{s-r}\}}\otimes\Sigma^{(1)}\left(L_s\right)\cdot\mathbf{1}_{\{L_s=L_{s-r}\}}. \tag{26}$$

The covariance matrix, from the above equation, $\text{cov}\left(\mathbf{X_s},\mathbf{X_{s-r}}|\mathbf{L};\Theta\right)$ has a separable structure between the spectral domain and temporal dimensions. It has a form analogous to that of what is shown in (4) through (6), which is intuitively appealing.

The number of parameters in the unpatterned interaction matrix coefficient is $N^2=N_1^2N_2^2$. On the other hand, the number of parameters for the separable interaction matrix coefficient is $N_1^2+N_2^2$, which has fewer parameters compared to its non-separable counterpart.

3.2.5 Separable mean structure

Likewise, we can also model the mean with a separable structure of the form

$$\mu(m)=\mu^{(2)}(m)\otimes\mu^{(1)}(m) \tag{27}$$

for $1\leq m\leq M$ where $\mu^{(1)}(m)\in\mathbb{R}^{N_1\times1}$ is the mean across the bands and $\mu^{(2)}(m)\in\mathbb{R}^{N_2\times1}$ is the mean across time. The number of parameters in the unpatterned mean vector is $N=N_1N_2$. On the other hand, the number of parameters for the separable mean vector is N_1+N_2 which has fewer number of parameters compared to its non-separable counterpart.

3.2.6 Hybrid separable structure

Finally, we can model the GMRF parameters as having a hybrid separability structure, that is, some of its parameters are separable while the rest are not. Hence, there are eight combinations to consider. As shown in Section 5.2, it is impossible to model a separable interaction matrix with a non-separabable matrix. This leave us six cases to consider in this study.

4. Estimation of thematic map parameters

The MPLE of φ is obtained by taking the derivative of $\log PL(\varphi)$ with respect to $\{a_m\}_{1\leq m\leq M}$ and $\{b_r\}_{r\in\mathcal{N}}$, then equating to zero (Li, 1995). Accordingly, the estimators are obtained numerically by solving the following set of simultaneous nonlinear equations:

$$\sum_{s\in\mathcal{S}}\frac{\exp\left(a_m+\sum_{r\in\mathcal{N}}b_r\cdot V\left(L_s=m,L_{s-r}\right)\right)}{\sum_{l=1}^M\exp\left(a_l+\sum_{r\in\mathcal{N}}b_r\cdot V\left(L_s=l,L_{s-r}\right)\right)}=\sum_{s\in\mathcal{S}}\mathbf{1}_{\{L_s=m\}}\quad\forall a_m,\ 1\leq m\leq M \tag{28}$$

$$\sum_{s\in\mathcal{S}}\frac{\sum_{l=1}^M\exp\left(a_l+\sum_{t\in\mathcal{N}}b_t\cdot V\left(L_s=l,L_{s-t}\right)\right)\cdot V\left(L_s=l,L_{s-r}\right)}{\sum_{l=1}^M\exp\left(a_l+\sum_{t\in\mathcal{N}}b_t\cdot V\left(L_s=l,L_{s-t}\right)\right)}=\sum_{s\in\mathcal{S}}V\left(L_s,L_{s-r}\right)\quad\forall b_r,\ \mathbf{r}\in\mathcal{N}. \tag{29}$$

5. Important MGMRF specifications

This section provides important characterizations enable us to derive the estimators of the GMRF parameters in the next section. We present a simple, yet powerful, method to derive the MPL estimators of the mean and the interaction matrix. Finally, new problems arise in estimating the multivariate observation GMRFs, which were not encountered in the univariate case, are discussed.

5.1 MPL-based method technique of deriving mean and the interaction matrix estimators

In this section, a method of deriving the MPL estimators for the mean and the vectorized interaction coefficients are presented regardless of separability. The MPL estimator of the interaction matrix coefficients can be derived by taking the matrix derivative of the log of the pseudo-likelihood function with respect to the interaction matrix coefficient or with respect to its vectorized version from the equivalence relation (Neudecker, 1969)

$$\frac{\partial f}{\partial \mathbf{X}} = \mathbf{P} \Leftrightarrow \frac{\partial f}{\partial vec(\mathbf{X})} = vec(\mathbf{P}) \tag{30}$$

where $f(\mathbf{X}) \in \mathbb{R}$, and $\mathbf{X}, \mathbf{P} \in \mathbb{R}^{m \times n}$. The latter expression is preferred, since it is easier to evaluate. The following proposition provides a simple way of deriving the MPL estimators, where the estimator is either the mean or the vectorized interaction matrix coefficient (Navarro, et. al., 2009).

Proposition 1   Let $\mathbf{\Phi}(m) \in \mathbb{R}^{q \times 1}$, $1 \le m \le M$ be a vector of parameters which is either the mean or the vectorized interaction matrix coefficient. Suppose that $\mathbf{X}_s$ can be expressed in the form

$$\mathbf{X}_s = \mathbf{P}_s - \mathbf{Q}_s \mathbf{\Phi}(L_s) \tag{31}$$

where $\mathbf{P}_s = \mathbf{P}_s(\mathbf{\Theta}|\mathbf{L}) \in \mathbb{R}^{N \times 1}, \mathbf{Q}_s = \mathbf{Q}_s(\mathbf{\Theta}|\mathbf{L}) \in \mathbb{R}^{N \times q}$ is independent of $\mathbf{\Phi}(L_s)$, and the covariance matrix $\mathbf{\Sigma}(m)$, $1 \le m \le M$ is known, then the MPL estimator for $\mathbf{\Phi}(m)$, $1 \le m \le M$ is obtained by solving the equation

$$\sum_{s \in \mathcal{S}(m)} \mathbf{Q}_s^T \mathbf{\Sigma}^{-1}(m) \mathbf{X}_s = \mathbf{0}_{q \times 1} . \tag{32}$$

Proof   From (7) and (9), the log pseudo-likelihood of the image random field conditional to the thematic map is given as

$$\log PL(\mathbf{\Theta}|\mathbf{L}) = -\frac{1}{2} \sum_{m=1}^{M} \sum_{s \in \mathcal{S}(m)} \left[N \log 2\pi + \log |\mathbf{\Sigma}(L_s)| + \mathbf{X}_s^T \mathbf{\Sigma}^{-1}(L_s) \mathbf{X}_s \right] . \tag{33}$$

Taking the gradient of the log pseudo-likelihood function in (33) with respect to $\mathbf{\Phi}(m)$ for $1 \le m \le M$, and equating to $\mathbf{0}_{q \times 1}$ yields

$$0_{q\times 1} = \frac{\partial}{\partial\Phi(m)}\log PL(\Theta|\mathbf{L}) = -\frac{1}{2}\sum_{l=1}^{M}\sum_{s\in\mathcal{S}(l)}\frac{\partial}{\partial\Phi(m)}\mathbf{X}_s^T\Sigma^{-1}(L_s)\mathbf{X}_s. \tag{34}$$

Since

$$\mathbf{X}_s^T\Sigma^{-1}(L_s)\mathbf{X}_s = (\mathbf{P}_s - \mathbf{Q}_s\Phi(L_s))^T\Sigma^{-1}(L_s)(\mathbf{P}_s - \mathbf{Q}_s\Phi(L_s))$$
$$= \mathbf{P}_s^T\Sigma^{-1}(L_s)\mathbf{P}_s - 2\mathbf{P}_s^T\Sigma^{-1}(L_s)\mathbf{Q}_s\Phi(L_s) + \Phi^T(L_s)\mathbf{Q}_s^T\Sigma^{-1}(L_s)\mathbf{Q}_s\Phi(L_s) \tag{35}$$

then taking the gradient in (34) with respect to Φ yields

$$\frac{\partial}{\partial\Phi}\mathbf{X}_s^T\Sigma^{-1}(L_s)\mathbf{X}_s = 2\mathbf{Q}_s^T\Sigma^{-1}(L_s)\mathbf{P}_s\mathbf{1}_{\{L_s=m\}} - 2\mathbf{Q}_s^T\Sigma^{-1}(L_s)\mathbf{Q}_s\Phi(L_s)\mathbf{1}_{\{L_s=m\}}$$
$$= 2\mathbf{Q}_s^T\Sigma^{-1}(L_s)(\mathbf{P}_s - \mathbf{Q}_s\Phi(L_s))\mathbf{1}_{\{L_s=m\}} \tag{36}$$
$$= 2\mathbf{Q}_s^T\Sigma^{-1}(L_s)\mathbf{X}_s\mathbf{1}_{\{L_s=m\}}.$$

Finally, substituting the result of (36) into (34) gives us the identity

$$0_{q\times 1} = \sum_{l=1}^{M}\sum_{s\in\mathcal{S}(l)}\mathbf{Q}_s^T\Sigma^{-1}(L_s)\mathbf{X}_s\mathbf{1}_{\{L_s=m\}} = \sum_{s\in\mathcal{S}(m)}\mathbf{Q}_s^T\Sigma^{-1}(m)\mathbf{X}_s. \tag{37}$$

5.2 Interaction matrix identities

From the covariance identity

$$\text{cov}(\mathbf{X}_s,\mathbf{X}_{s-r}|\mathbf{L};\Theta) = \text{cov}^T(\mathbf{X}_{s-r},\mathbf{X}_s|\mathbf{L};\Theta) \tag{38}$$

(Ravishanker and Dey, 2002), from (5), we obtain the following relationship:

$$\theta_{-r}(L_s) = \Sigma(L_s)\theta_r^T(L_s)\Sigma^{-1}(L_s). \tag{39}$$

One consequence of this result is that $\mathbf{X}_s$ can be written as follows:

$$\mathbf{X}_s = (\mathbf{Y}_s - \mu(L_s)) - \sum_{r\in\mathcal{N}_s}\left[\theta_r(L_s)\mathbf{1}_{\{L_s=L_{s-r}\}}(\mathbf{Y}_{s-r} - \mu(L_s)) + \Sigma(L_s)\theta_r^T(L_s)\Sigma^{-1}(L_s)\mathbf{1}_{\{L_s=L_{s+r}\}}(\mathbf{Y}_{s+r} - \mu(L_s))\right] \tag{40}$$

where $\mathcal{N}_s$, a subset of $\mathcal{N}$ which represents the symmetric neighborhood set (Kashyap and Chellappa, 1983), is defined as follows: $\mathbf{r}\in\mathcal{N}_s \Rightarrow -\mathbf{r}\notin\mathcal{N}_s$ and $\mathcal{N}=\{\mathbf{r}\in\mathcal{N}_s\cup -\mathbf{r}\in\mathcal{N}_s\}$.

Another consequence of (39) are the specifications of the interaction matrices in the separable case. If the interaction matrices are modeled as separable, then by (39), we obtain

$$\theta_{-r}(m) = \theta_{-r}^{(2)}(m)\otimes\theta_{-r}^{(1)}(m) = \Sigma(m)(\theta_r^{(2)}(m)\otimes\theta_r^{(1)}(m))^T\Sigma^{-1}(m) = \Sigma(m)\theta_r^T(m)\Sigma^{-1}(m) \tag{41}$$

for $1\le m\le M$. The RHS of (40) can be made separable if $\Sigma(m)$ is also separable. Hence,

$$\theta_{-r}^{(2)}(m) \otimes \theta_{-r}^{(1)}(m) = \left(\Sigma^{(2)}(m) \otimes \Sigma^{(1)}(m)\right)\left(\theta_r^{(2)}(m) \otimes \theta_r^{(1)}(m)\right)^T \left(\Sigma^{(2)}(m) \otimes \Sigma^{(1)}(m)\right)^{-1}$$

$$= \left(\Sigma^{(2)}(m) \otimes \Sigma^{(1)}(m)\right)\left(\theta_r^{(2)T}(m) \otimes \theta_r^{(1)T}(m)\right)\left(\left(\Sigma^{(2)}(m)\right)^{-1} \otimes \left(\Sigma^{(1)}(m)\right)^{-1}\right) \quad (42)$$

$$= \Sigma^{(2)}(m)\theta_r^{(2)T}(m)\left(\Sigma^{(2)}(m)\right)^{-1} \otimes \Sigma^{(1)}(m)\theta_r^{(1)T}(m)\left(\Sigma^{(1)}(m)\right)^{-1}.$$

The identification of $\theta_{-r}(m)$ is completely specified from (39) if we take

$$\theta_{-r}^{(1)}(m) = \Sigma^{(1)}(m)\theta_r^{(1)T}(m)\left(\Sigma^{(1)}(m)\right)^{-1} \quad (43)$$

$$\theta_{-r}^{(2)}(m) = \Sigma^{(2)}(m)\theta_r^{(2)T}(m)\left(\Sigma^{(2)}(m)\right)^{-1}, \quad (44)$$

which is analogous to the relation in (39).

By considering the hybrid separability cases which involve a separable interaction matrix and a non-separable covariance matrix, the expression $\Sigma(m)\theta_r^T(m)\Sigma^{-1}(m)$ is not separable, in general. This implies that $\theta_{-r}(m)$ cannot be expressed in the form $\theta_{-r}(m) = \theta_{-r}^{(2)}(m) \otimes \theta_{-r}^{(1)}(m)$ for $r \in \mathcal{N}_S, 1 \le m \le M$ and thus these cases are not possible.

6. GMRF parameter estimation

This section proposes an estimation procedure for the GMRF parameters for both separable and non-separable cases based on the MPL.

6.1 Mean parameter estimation

Proposition 2   Assume that the interaction matrix coefficients $\theta_r(m)$ for $r \in \mathcal{N}$, $1 \le m \le M$ and the covariance matrices $\Sigma(m)$ for $1 \le m \le M$ are known. Then the mean parameters are estimated as follows:

a.   Non-Separable Case:

$$\hat{\mu}(m) = \left[\sum_{s \in S(m)}\left(\mathbf{I}_N - \sum_{r \in \mathcal{N}}\theta_r(m)\mathbf{1}_{\{L_{s-r}=m\}}\right)^T \Sigma^{-1}(m)\left(\mathbf{I}_N - \sum_{r \in \mathcal{N}}\theta_r(m)\mathbf{1}_{\{L_{s-r}=m\}}\right)\right]^{-1} \cdot$$

$$\quad (45)$$

$$\left[\sum_{s \in S(m)}\left(\mathbf{I}_N - \sum_{r \in \mathcal{N}}\theta_r(m)\mathbf{1}_{\{L_{s-r}=m\}}\right)^T \Sigma^{-1}(m)\left(\mathbf{Y}_s - \sum_{r \in \mathcal{N}}\theta_r(m)\mathbf{1}_{\{L_{s-r}=m\}}\mathbf{Y}_{s-r}\right)\right]$$

for $1 \le m \le M$.

b.   Separable Case:

In addition, if we assume the following for $1 \le m \le M$:

- $\boldsymbol{\mu}^{(1)}(m)$ is estimated, given that $\boldsymbol{\mu}^{(2)}(m)$ is known
- $\boldsymbol{\mu}^{(2)}(m)$ is estimated, given that $\boldsymbol{\mu}^{(1)}(m)$ is known.

Thus

$$\hat{\boldsymbol{\mu}}^{(1)}(m)$$

$$= \left[\sum_{s \in \mathcal{S}(m)} \left(\boldsymbol{\mu}^{(2)}(m) \otimes \mathbf{I}_{N_1} \right)^T \left(\mathbf{I}_N - \sum_{r \in \mathcal{N}} \boldsymbol{\theta}_r(m) \mathbf{1}_{\{L_{s-r}=m\}} \right)^T \boldsymbol{\Sigma}^{-1}(m) \left(\mathbf{I}_N - \sum_{r \in \mathcal{N}} \boldsymbol{\theta}_r(m) \mathbf{1}_{\{L_{s-r}=m\}} \right) \left(\boldsymbol{\mu}^{(2)}(m) \otimes \mathbf{I}_{N_1} \right) \right]^{-1} \cdot \quad (46)$$

$$\left[\sum_{s \in \mathcal{S}(m)} \left(\boldsymbol{\mu}^{(2)}(m) \otimes \mathbf{I}_{N_1} \right)^T \left(\mathbf{I}_N - \sum_{r \in \mathcal{N}} \boldsymbol{\theta}_r(m) \mathbf{1}_{\{L_{s-r}=m\}} \right)^T \boldsymbol{\Sigma}^{-1}(m) \left(\mathbf{Y}_s - \sum_{r \in \mathcal{N}} \boldsymbol{\theta}_r(m) \mathbf{1}_{\{L_{s-r}=m\}} \mathbf{Y}_{s-r} \right) \right]$$

$$\hat{\boldsymbol{\mu}}^{(2)}(m)$$

$$= \left[\sum_{s \in \mathcal{S}(m)} \left(\mathbf{I}_{N_2} \otimes \boldsymbol{\mu}^{(1)}(m) \right)^T \left(\mathbf{I}_N - \sum_{r \in \mathcal{N}} \boldsymbol{\theta}_r(m) \mathbf{1}_{\{L_{s-r}=m\}} \right)^T \boldsymbol{\Sigma}^{-1}(m) \left(\mathbf{I}_N - \sum_{r \in \mathcal{N}} \boldsymbol{\theta}_r(m) \mathbf{1}_{\{L_{s-r}=m\}} \right) \left(\mathbf{I}_{N_2} \otimes \boldsymbol{\mu}^{(1)}(m) \right) \right]^{-1} \cdot \quad (47)$$

$$\left[\sum_{s \in \mathcal{S}(m)} \left(\mathbf{I}_{N_2} \otimes \boldsymbol{\mu}^{(1)}(m) \right)^T \left(\mathbf{I}_N - \sum_{r \in \mathcal{N}} \boldsymbol{\theta}_r(m) \mathbf{1}_{\{L_{s-r}=m\}} \right)^T \boldsymbol{\Sigma}^{-1}(m) \left(\mathbf{Y}_s - \sum_{r \in \mathcal{N}} \boldsymbol{\theta}_r(m) \mathbf{1}_{\{L_{s-r}=m\}} \mathbf{Y}_{s-r} \right) \right]$$

for $1 \le m \le M$.

Proof

a. The proof for the non-separable case is derived by applying Proposition 1 (Navarro, et. al., 2009).

b. From (3), $\mathbf{X}_s$ can be written as follows:

$$\mathbf{X}_s = \left(\mathbf{Y}_s - \sum_{r \in \mathcal{N}} \boldsymbol{\theta}_r(L_s) \mathbf{1}_{\{L_{s-r}=L_s\}} \mathbf{Y}_{s-r} \right) - \left(\mathbf{I}_N - \sum_{r \in \mathcal{N}} \boldsymbol{\theta}_r(L_s) \mathbf{1}_{\{L_{s-r}=L_s\}} \right) \boldsymbol{\mu}(L_s). \quad (48)$$

For the separable case, the mean can be written as follows:

$$\begin{aligned}
\boldsymbol{\mu}(m) &= \boldsymbol{\mu}^{(2)}(m) \otimes \boldsymbol{\mu}^{(1)}(m) \\
&= \left(\boldsymbol{\mu}^{(2)}(m) \otimes \mathbf{I}_{N_1} \right) \left(1 \otimes \boldsymbol{\mu}^{(1)}(m) \right) = \left(\boldsymbol{\mu}^{(2)}(m) \otimes \mathbf{I}_{N_1} \right) \boldsymbol{\mu}^{(1)}(m) \\
&= \left(\mathbf{I}_{N_2} \otimes \boldsymbol{\mu}^{(1)}(m) \right) \left(\boldsymbol{\mu}^{(2)}(m) \otimes 1 \right) = \left(\mathbf{I}_{N_2} \otimes \boldsymbol{\mu}^{(1)}(m) \right) \boldsymbol{\mu}^{(2)}(m).
\end{aligned} \quad (49)$$

Plugging the results of (49) into (48) yields

$$\begin{aligned}
\mathbf{X}_s &= \left(\mathbf{Y}_s - \sum_{r \in \mathcal{N}} \boldsymbol{\theta}_r(L_s) \mathbf{1}_{\{L_{s-r}=L_s\}} \mathbf{Y}_{s-r} \right) - \left(\mathbf{I}_N - \sum_{r \in \mathcal{N}} \boldsymbol{\theta}_r(L_s) \mathbf{1}_{\{L_{s-r}=L_s\}} \right) \left(\boldsymbol{\mu}^{(2)}(L_s) \otimes \mathbf{I}_{N_1} \right) \boldsymbol{\mu}^{(1)}(L_s) \\
&= \left(\mathbf{Y}_s - \sum_{r \in \mathcal{N}} \boldsymbol{\theta}_r(L_s) \mathbf{1}_{\{L_{s-r}=L_s\}} \mathbf{Y}_{s-r} \right) - \left(\mathbf{I}_N - \sum_{r \in \mathcal{N}} \boldsymbol{\theta}_r(L_s) \mathbf{1}_{\{L_{s-r}=L_s\}} \right) \left(\mathbf{I}_{N_2} \otimes \boldsymbol{\mu}^{(1)}(L_s) \right) \boldsymbol{\mu}^{(2)}(L_s).
\end{aligned} \quad (50)$$

$$\left(1^{\circ}\right) \qquad \Phi(m) = \mu^{(1)}(m), \ 1 \le m \le M$$

For this case, we recognize the following from (50):

$$Q_s = \left(I_N - \sum_{r \in \mathcal{N}} \theta_r(L_s) \mathbf{1}_{\{L_{s-r} = L_s\}}\right)\left(\mu^{(2)}(L_s) \otimes I_{N_1}\right). \qquad (51)$$

By applying Preposition 1 and rearranging terms, we obtain (46).

$$\left(2^{\circ}\right) \qquad \Phi(m) = \mu^{(1)}(m), \ 1 \le m \le M$$

For this case from (50), we recognize

$$Q_s = \left(I_N - \sum_{r \in \mathcal{N}} \theta_r(L_s) \mathbf{1}_{\{L_{s-r} = L_s\}}\right)\left(I_{N_2} \otimes \mu^{(1)}(L_s)\right). \qquad (52)$$

by applying Preposition 1 and rearranging terms, we obtain (47).

6.2 Interaction matrix parameter estimation

Proposition 3   Assume that the mean vectors $\mu(m)$ for $1 \le m \le M$ and the covariance matrices $\Sigma(m)$ for $1 \le m \le M$ are known, then interaction matrix parameters are estimated by solving the simultaneous linear equations given as follows:

a.   Non-Separable Case:

$$\mathbf{H}(m)\Psi(m) = \Gamma(m) \qquad (53)$$

where

$$\mathbf{H}(m) = row\left\{col\left(\sum_{s \in \mathcal{S}(m)} \mathbf{A}_{s,t}(m)\Sigma^{-1}(m)\mathbf{A}_{s,r}^T(m), \ r \in \mathcal{N}_S\right), \ t \in \mathcal{N}_S\right\} \qquad (54)$$

$$\Gamma(m) = row\left(\sum_{s \in \mathcal{S}(m)} \mathbf{A}_{s,t}(m)\Sigma^{-1}(m)\left(Y_s - \mu(m)\right), \ t \in \mathcal{N}_S\right) \qquad (55)$$

$$\Psi(m) = row\left(vec\left(\hat{\theta}_r(m)\right), \ r \in \mathcal{N}_S\right) \qquad (56)$$

and

$$\mathbf{A}_{s,r}(m) = \left(\left(Y_{s-r} - \mu(m)\right)\mathbf{1}_{\{L_{s-r} = m\}} \otimes I_N\right) + K_{N,N}\left(\Sigma^{-1}(m) \otimes \Sigma(m)\right)\left(\left(Y_{s+r} - \mu(m)\right)\mathbf{1}_{\{L_{s+r} = m\}} \otimes I_N\right). \qquad (57)$$

From the invariance property of the MPL, the complete set of non-separable interaction matrix estimators is estimated as follows:

$$\hat{\boldsymbol{\theta}}_{\mathbf{r}}(m) = reshape\left(vec\left(\hat{\boldsymbol{\theta}}_{\mathbf{r}}(m)\right), N, N\right) \tag{58}$$

$$\hat{\boldsymbol{\theta}}_{-\mathbf{r}}(m) = \boldsymbol{\Sigma}(m)\hat{\boldsymbol{\theta}}_{\mathbf{r}}^{T}(m)\left(\boldsymbol{\Sigma}(m)\right)^{-1} \tag{59}$$

for $\mathbf{r} \in \mathcal{N}_s$, $1 \le m \le M$.

b.   Separable Case:

In addition, if we assume the following for $\mathbf{r} \in \mathcal{N}_s$ and $1 \le m \le M$:

- $\boldsymbol{\theta}_{r}^{(1)}(m)$ is estimated, given that $\boldsymbol{\theta}_{r}^{(2)}(m)$ is known
- $\boldsymbol{\theta}_{r}^{(2)}(m)$ is estimated, given that $\boldsymbol{\theta}_{r}^{(1)}(m)$ is known

then

$$\mathbf{H}^{(k)}(m)\boldsymbol{\Psi}^{(k)}(m) = \boldsymbol{\Gamma}^{(k)}(m) \tag{60}$$

where

$$\mathbf{H}^{(k)}(m) = row\left\{col\left(\sum_{s \in \mathcal{S}(m)} \mathbf{A}_{s,t}^{(k)}(m)\boldsymbol{\Sigma}^{-1}(m)\mathbf{A}_{s,r}^{(k)T}(m), \ \mathbf{r} \in \mathcal{N}_S\right), \ \mathbf{t} \in \mathcal{N}_S\right\} \tag{61}$$

$$\boldsymbol{\Gamma}^{(k)}(m) = row\left(\sum_{s \in \mathcal{S}(m)} \mathbf{A}_{s,t}^{(k)}(m)\boldsymbol{\Sigma}^{-1}(m)\left(\mathbf{Y}_s - \boldsymbol{\mu}(m)\right), \ \mathbf{t} \in \mathcal{N}_S\right) \tag{62}$$

$$\boldsymbol{\Psi}^{(k)}(m) = row\left(vec\left(\hat{\boldsymbol{\theta}}_{\mathbf{r}}^{(k)}(m)\right), \ \mathbf{r} \in \mathcal{N}_S\right) \tag{63}$$

for $1 \le k \le 2$ and

$$\mathbf{A}_{s,r}^{(1)}(m) = \left(vec\left(\boldsymbol{\theta}_{\mathbf{r}}^{(2)}(m)\right) \otimes \mathbf{I}_{N_1}\right)^{T}\left(\mathbf{I}_{N_2} \otimes \mathbf{K}_{N_1,N_2} \otimes \mathbf{I}_{N_1}\right)^{T}\mathbf{A}_{s,r}(m) \tag{64}$$

$$\mathbf{A}_{s,r}^{(2)}(m) = \left(\mathbf{I}_{N_2} \otimes vec\left(\boldsymbol{\theta}_{\mathbf{r}}^{(1)}(m)\right)\right)^{T}\left(\mathbf{I}_{N_2} \otimes \mathbf{K}_{N_1,N_2} \otimes \mathbf{I}_{N_1}\right)^{T}\mathbf{A}_{s,r}(m). \tag{65}$$

From the invariance property of the MPL, the complete set of separable interaction matrix estimators is estimated as follows for $\mathbf{r} \in \mathcal{N}_s$, $1 \le m \le M$, $1 \le k \le 2$:

$$\hat{\boldsymbol{\theta}}_{\mathbf{r}}^{(k)}(m) = reshape\left(vec\left(\hat{\boldsymbol{\theta}}_{\mathbf{r}}^{(k)}(m)\right), N_k, N_k\right) \tag{66}$$

$$\hat{\boldsymbol{\theta}}_{\mathbf{r}}^{(k)}(m) = \boldsymbol{\Sigma}^{(k)}(m)\hat{\boldsymbol{\theta}}_{\mathbf{r}}^{(k)T}(m)\left(\boldsymbol{\Sigma}^{(k)}(m)\right)^{-1} \tag{67}$$

and also

$$\hat{\boldsymbol{\theta}}_{\mathbf{r}}(m) = \hat{\boldsymbol{\theta}}_{\mathbf{r}}^{(2)}(m) \otimes \hat{\boldsymbol{\theta}}_{\mathbf{r}}^{(1)}(m) \tag{68}$$

for $\mathbf{r} \in \mathcal{N}_s$ and $1 \le m \le M$.

Proof

a. The proof for the non-separable case is derived by applying Proposition 1 (Navarro, et. al., 2009).
b. From (3), $\mathbf{X_s}$ can be written as

$$\mathbf{X_s} = vec(\mathbf{X_s}) = \left(\mathbf{Y_s} - \boldsymbol{\mu}(L_s)\right) - \sum_{r \in \mathcal{N}_S} \mathbf{A}_{s,r}^T(L_s) vec\left(\boldsymbol{\theta}_{\mathbf{r}}(L_s)\right). \tag{69}$$

The above expression can also be written using the following matrix identities (Magnus and Neudecker, 1999)

$$vec(\mathbf{ABC}) = \left(\mathbf{C}^T \otimes \mathbf{A}\right) vec(\mathbf{B}) \tag{70}$$

where $\mathbf{A} \in \mathbb{R}^{m \times n}$, $\mathbf{B} \in \mathbb{R}^{n \times p}$, and $\mathbf{C} \in \mathbb{R}^{p \times q}$.

$$(\mathbf{A} \otimes \mathbf{B})^T = \mathbf{A}^T \otimes \mathbf{B}^T \tag{71}$$

$$vec(\mathbf{A}^T) = \mathbf{K}_{m,n} vec(\mathbf{A}) \tag{72}$$

where $\mathbf{A} \in \mathbb{R}^{m \times n}$. In addition, from the identity (Magnus and Neudecker, 1999)

$$vec(\mathbf{A} \otimes \mathbf{B}) = \left(\mathbf{I}_n \otimes \mathbf{K}_{q,m} \otimes \mathbf{I}_p\right) \cdot \left(vec(\mathbf{A}) \otimes vec(\mathbf{B})\right) \tag{73}$$

where $\mathbf{A} \in \mathbb{R}^{m \times n}$ and $\mathbf{B} \in \mathbb{R}^{p \times q}$, it follows that

$$\begin{aligned} vec\left(\boldsymbol{\theta}_{\mathbf{r}}(m)\right) &= vec\left(\boldsymbol{\theta}_{\mathbf{r}}^{(2)}(m) \otimes \boldsymbol{\theta}_{\mathbf{r}}^{(1)}(m)\right) \\ &= \left(\mathbf{I}_{N_2} \otimes \mathbf{K}_{N_1,N_2} \otimes \mathbf{I}_{N_1}\right) \cdot \left(vec\left(\boldsymbol{\theta}_{\mathbf{r}}^{(2)}(m)\right) \otimes vec\left(\boldsymbol{\theta}_{\mathbf{r}}^{(1)}(m)\right)\right). \end{aligned} \tag{74}$$

Furthermore, since

$$\begin{aligned} &\left(vec\left(\boldsymbol{\theta}_{\mathbf{r}}^{(2)}(m)\right) \otimes vec\left(\boldsymbol{\theta}_{\mathbf{r}}^{(1)}(m)\right)\right) \\ &= \left(vec\left(\boldsymbol{\theta}_{\mathbf{r}}^{(2)}(m)\right) \otimes \mathbf{I}_{N_1}\right)\left(1 \otimes vec\left(\boldsymbol{\theta}_{\mathbf{r}}^{(1)}(m)\right)\right) = \left(vec\left(\boldsymbol{\theta}_{\mathbf{r}}^{(2)}(m)\right) \otimes \mathbf{I}_{N_1}\right) vec\left(\boldsymbol{\theta}_{\mathbf{r}}^{(1)}(m)\right) \\ &= \left(\mathbf{I}_{N_2} \otimes vec\left(\boldsymbol{\theta}_{\mathbf{r}}^{(1)}(m)\right)\right)\left(vec\left(\boldsymbol{\theta}_{\mathbf{r}}^{(2)}(m)\right) \otimes 1\right) = \left(\mathbf{I}_{N_2} \otimes vec\left(\boldsymbol{\theta}_{\mathbf{r}}^{(1)}(m)\right)\right) vec\left(\boldsymbol{\theta}_{\mathbf{r}}^{(2)}(m)\right) \end{aligned} \tag{75}$$

then,

$$\begin{aligned} vec\left(\boldsymbol{\theta}_{\mathbf{r}}(m)\right) &= \left(\mathbf{I}_{N_2} \otimes \mathbf{K}_{N_1,N_2} \otimes \mathbf{I}_{N_1}\right) \cdot \left(vec\left(\boldsymbol{\theta}_{\mathbf{r}}^{(2)}(m)\right) \otimes \mathbf{I}_{N_1}\right) vec\left(\boldsymbol{\theta}_{\mathbf{r}}^{(1)}(m)\right) \\ &= \left(\mathbf{I}_{N_2} \otimes \mathbf{K}_{N_1,N_2} \otimes \mathbf{I}_{N_1}\right) \cdot \left(\mathbf{I}_{N_2} \otimes vec\left(\boldsymbol{\theta}_{\mathbf{r}}^{(1)}(m)\right)\right) vec\left(\boldsymbol{\theta}_{\mathbf{r}}^{(2)}(m)\right). \end{aligned} \tag{76}$$

Plugging the results of (76) into (69) yields

$$
\begin{aligned}
\mathbf{X_s} &= \left(\mathbf{Y_s} - \boldsymbol{\mu}(L_s)\right) - \sum_{r \in \mathcal{N}_S} \mathbf{A}_{s,r}^{(1)T}(L_s) vec\left(\boldsymbol{\theta}_r^{(1)}(L_s)\right) \\
&= \left(\mathbf{Y_s} - \boldsymbol{\mu}(L_s)\right) - \sum_{r \in \mathcal{N}_S} \mathbf{A}_{s,r}^{(2)T}(L_s) vec\left(\boldsymbol{\theta}_r^{(2)}(L_s)\right)
\end{aligned}
\tag{77}
$$

$$
\left(1^{\cdot}\right) \quad \boldsymbol{\Phi}(m) = vec\left(\boldsymbol{\theta}_t^{(1)}(m)\right), \ t \in \mathcal{N}_s, \ 1 \le m \le M
$$

For this case, we recognize from (77),

$$
\mathbf{Q_s} = \mathbf{A}_{s,t}^{(1)T}(m). \tag{78}
$$

By applying Preposition 1 and rearranging terms, we obtain the following expression

$$
\sum_{r \in \mathcal{N}_S} \sum_{s \in \mathcal{S}(m)} \mathbf{A}_{s,t}^{(1)}(m)\boldsymbol{\Sigma}^{-1}(m)\mathbf{A}_{s,r}^{(1)T}(m) vec\left(\boldsymbol{\theta}_r^{(1)}(m)\right) = \sum_{s \in \mathcal{S}(m)} \mathbf{A}_{s,t}^{(1)}(m)\boldsymbol{\Sigma}^{-1}(m)\left(\mathbf{Y_s} - \boldsymbol{\mu}(m)\right). \tag{79}
$$

By aggregating the equation in (79) for $t \in \mathcal{N}_S$, the interaction matrix coefficients are estimated by solving the simultaneous linear equations in (60) for $k = 1$.

$$
\left(2^{\cdot}\right) \quad \boldsymbol{\Phi}(m) = vec\left(\boldsymbol{\theta}_t^{(2)}(m)\right), \ t \in \mathcal{N}_s, \ 1 \le m \le M
$$

For this case, we recognize from (77)

$$
\mathbf{Q_s} = \mathbf{A}_{s,t}^{(2)T}(m). \tag{80}
$$

By applying Preposition 1 and rearranging terms, we obtain the following expression

$$
\sum_{r \in \mathcal{N}_S} \sum_{s \in \mathcal{S}(m)} \mathbf{A}_{s,t}^{(2)}(m)\boldsymbol{\Sigma}^{-1}(m)\mathbf{A}_{s,r}^{(2)T}(m) vec\left(\boldsymbol{\theta}_r^{(2)}(m)\right) = \sum_{s \in \mathcal{S}(m)} \mathbf{A}_{s,t}^{(2)}(m)\boldsymbol{\Sigma}^{-1}(m)\left(\mathbf{Y_s} - \boldsymbol{\mu}(m)\right). \tag{81}
$$

By aggregating the equations in (79) for $t \in \mathcal{N}_S$, the interaction matrix coefficients are estimated by solving the simultaneous linear equations in (60) for $k = 2$.

6.3 Covariance matrix parameter estimation

Since $\mathbf{X}_s$ is dependent on a covariance matrix in finding the MPL estimator of $\boldsymbol{\Sigma}(m)$, for all $1 \le m \le M$ is cumbersome to derive. As an alternative, we estimate the covariance matrix as the sample covariance matrix given that the mean vectors $\boldsymbol{\mu}(m)$ for $1 \le m \le M$ and the interaction matrix coefficients $\boldsymbol{\theta}_r(m)$, for $r \in \mathcal{N}, 1 \le m \le M$ are known, then the covariance matrix parameters are estimated as follows:

a.   Non-Separable Case:

$$
\hat{\boldsymbol{\Sigma}}(m) = \frac{1}{r(m)} \sum_{s \in \mathcal{S}(m)} \mathbf{X_s}\mathbf{X_s}^T \tag{82}
$$

b.   Separable Case:

In addition, if we assume the following for $1 \leq m \leq M$:

- $\Sigma^{(1)}(m)$ is estimated, given that $\Sigma^{(2)}(m)$ is known
- $\Sigma^{(2)}(m)$ is estimated, given that $\Sigma^{(1)}(m)$ is known

then

$$\hat{\Sigma}^{(1)}(m) = \frac{1}{r(m)N_2} \sum_{s \in \mathcal{S}(m)} \mathbf{X}_s^{\#} \left(\hat{\Sigma}^{(2)}(m) \right)^{-1} \mathbf{X}_s^{\#T} \qquad (83)$$

$$\hat{\Sigma}^{(2)}(m) = \frac{1}{r(m)N_1} \sum_{s \in \mathcal{S}(m)} \mathbf{X}_s^{\#T} \left(\Sigma^{(1)}(m) \right)^{-1} \mathbf{X}_s^{\#}. \qquad (84)$$

The above estimators are not in their closed form. The estimators can be solved iteratively using the flip-flop algorithm (Dutilleul, 1999).

7. Data preparation

The multispectral and multitemporal satellite image under consideration is the 'Butuan' image acquired from the LANDSAT TM. The image shows the scenery of Butuan City and its surroundings in Northeastern Mindanao, Philippines. It consists of six spectral bands and four temporal slots with a dynamic range of 8 bits. The images were captured chronologically on the following dates: August 1, 1992, August 7, 2000, May 22, 2001, and December 3, 2002. The images were radiometrically corrected, geometrically co-registered with each other, and have been resized to 600 x 800 pixels. The image in Fig. 1 is a gray-scaled RGB realization captured on May 22, 2001.

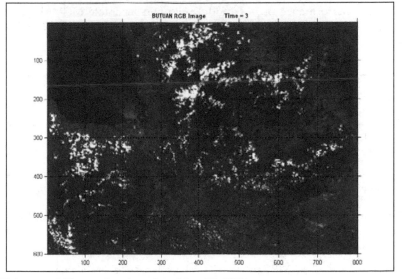

Fig. 1. RGB image of 'Butuan' captured on May 22, 2001.

The thematic classes were established by employing the k-means algorithm (Richards and Jia, 2006). The thematic classes were identified and their mean reflectance vector form the training data are shown in Table 1.

M	Thematic Class	Landsat TM Band Number					
		1	2	3	4	5	7
1	Thick Vegetation	62	48	33	91	69	29
2	Sparse Vegetation	70	58	43	99	83	37
3	Built Up Areas	77	63	54	75	78	41
4	Body of Water	72	41	29	12	13	11
5	Thin Clouds	104	84	76	88	85	53
6	Thick Clouds	197	190	190	144	167	115

Table 1. Average reflectances from the training data.

Training and verification sites were obtained from a random sample of 1200 sites. The first-order neighborhood system in the MRF modeling of the thematic map and the image were used.

8. Discussion

8.1 Non-separable case

The classification performance of our model with non-separable MGMRF parameters, as compared to the GSC, Hazel's, and Rellier's models are presented in Table 2.

Model	Accuracy
GSC	55.3%
Hazel's GMRF	45.6%
Rellier's GMRF	83.1%
Our Model	84.3%

Table 2. Classification Accuracy of Different MGMRF models.

The GSC model has a low accuracy compared to the remaining MGMRF models. It substantiates that Markov dependence would yield a better accuracy to the thematic map classification than to the site independence model.

It is noticeable that Hazel's GMRF presents a relatively poor classification accuracy which is attributed to the bilateral symmetry imposed into the interaction matrices, that is,

$$\theta_r(L_s) = \theta_{-r}(L_s) \tag{85}$$

(Hazel, 2000) which in general, does not hold the multivariate case. This relation, however, holds in the univariate case (Kashyap and Chellappa, 1983) as well as the Rellier's GMRF.

On the other hand, anisotropic models, such as Rellier's GMRF, and our model exhibited a substantially better classification performance as compared to the GSC. Since the covariance matrix estimators used a sub-optimal alternative, some slight performance degradation has resulted.

8.2 Hybrid separable case

Denote S_μ, S_θ, and S_Σ to be the separable indicators for the mean, interaction matrix, and covariance matrix, respectively.

8.2.1 Hybrid separable GSC model

Since the GSC model is a degenerate form of our MGMRF with zero interaction matrices, the separability structure of the mean and covariance matrices are examined. The results are presented in Table 3 showed that no improvement in the classification performance, regardless of separability of the parameters.

S_Σ	S_μ	Accuracy
0	0	55.3%
0	1	54.2%
0	0	54.3%
1	1	54.1%

Table 3. Classification Accuracy of Hybrid Separable GSC models.

8.2.2 Hybrid separable anisotropic GMRF model

The hybrid separable anisotropic MGMRF shows the separability of the covariance matrix has a slight improvement in performance over a non-separable spectro-temporal observation. As discussed in Section 5.2, the hybrid separable model with separable interaction matrix, together with a non-separable matrix, were excluded in the model performance as these modes are not possible. The classification accuracy is presented in Table 4.

S_Σ	S_θ	S_μ	Accuracy
0	0	0	84.3%
0	0	1	84.6%
0	1	0	
0	1	1	

S_Σ	S_θ	S_μ	Accuracy
1	0	0	84.5%
1	0	1	86.6%
1	1	0	83.8%
1	1	1	86.2%

Table 4. Classification Accuracy of Hybrid Separable Anisotropic MGMRF models.

8.3 Thematic maps

Some of the thematic map labels are presented in Figs. 2 to 4, based on the May 22, 2001 satellite image. For clarity of visual presentation, thematic map labels were based on the gray-scaled average RGB reflectance of the training data.

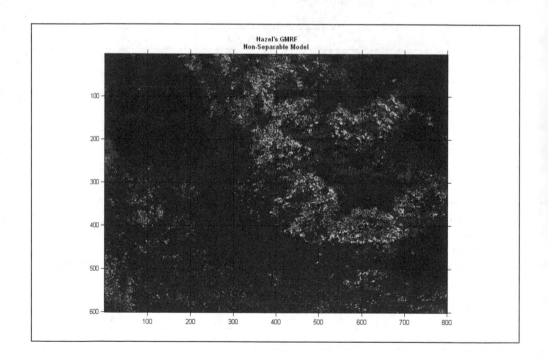

Fig. 2. Thematic Map – Hazel's MGMRF

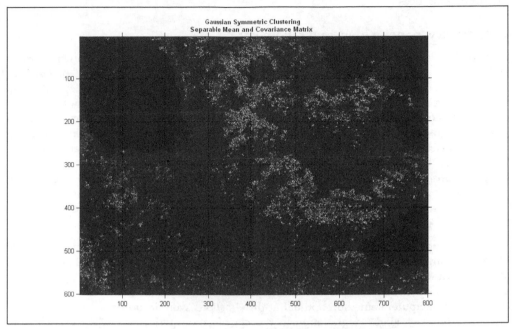

Fig. 3. Thematic Map – GSC with separable mean and covariance matrix

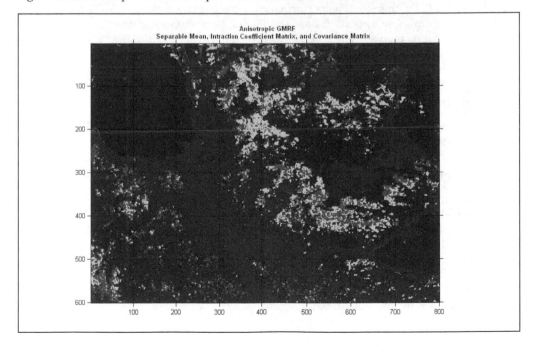

Fig. 4. Thematic Map – Anisotropic MGMRF separable mean, interaction matrices, and covaraince matrices

9. Summary, conclusions, and recommendations

This study presents a parameter estimation procedure based on the MPL for an anisotropic MGMRF with hybrid-separable parameters. Although the MGMRF is a natural extension of its univariate counterpart, the interaction matrix relationship is, in general, dependent on the covariance matrix. In an effort to make the estimation and classification procedure more tractable to compute, some sub-optimal approximations were incorporated. This resulted in a slight degradation in the classification performance. The classification performance based on our model performed well when compared to the GSC model and Hazel's MGMRF. Nonetheless, its performance is comparable to the Rellier's MGMRF. Moreover, for spectro-temporal observations, the separability of the interaction matrix as well as the covariance matrix improved the classification performance. Computational capabilities are foreseen to further advance in the near future following the improvement of numerical estimation and classification procedures.

This study presents a parameter estimation procedure based on the MPL for anisotropic MGMRF with hybrid-separable parameters. Although the MGMRF is a natural extension of its univariate counterpart, the interaction matrix relationship is, in general, dependent on the covariance matrix. In an effort to make the estimation and classification procedure more tractable to compute, some sub-optimal approximations were incorporated in the process. This resulted in a slight degradation in the classification performance. The classification performance, based on our model, has performed well, as compared to the GSC model and Hazel's MGMRF. Furthermore, its performance is comparable to Rellier's MGMRF. In terms of spectro-temporal observations, the separability of the covariance matrix has improved the classification performance. This study can be improved even more with numerical estimation and classification procedure as computational capabilities. This is foreseen to further advance in the near future.

10. Acknowledgment

We acknowledge the invaluable support of extended by the Statistical Training and Research Center of the Philippine Statistical System.

11. References

Aarts, E. and Korts, J. (1987). *Simulated Annealing and Boltzmann Machines*, Wiley, ISBN 978-047-1921-46-2, New York

Arnold, S. F. (1981). *The Theory of Linear Models and Multivariate Analysis*, Wiley, ISBN 978-047-1050-65-0, New York

Besag, J. (1986). On the Statistical Analysis of Dirty Pictures (with discussions). *Journal of Royal Statistical Society B*. Vol. 48, No. 3, pp. 259-302, ISSN 0035-9246

Campbell, N. A. and Kiiveri, H. T. (1988). Spectral-Temporal Indices for Discrimination. *Applied Statistics*, Vol. 37, No. 1, pp. 51-62, ISSN 0035-9254

Casella, G. & Berger, R. L. (2002). *Statistical Inference 2nd ed.*, Wadswoth Group, ISBN 978-053-4243-12-8, Pacific Grove, CA

Dutilleul, P. (1999). The MLE Algorithm for the Matrix Normal Distribution. *Journal of Statistical Computation and Simulation*, Vol. 64, No. 2 , ISSN 0094-9655

Fuentes, M. (2006). Testing for Separability of Spatial-Temporal Covariance Functions. *Journal of Statistical Planning and Inference*, Vol. 136, pp. 447-466, ISSN 0378-3758

Geman, S. & Graffigne, C. (1987). Markov Random Field Models and Their Applications to Computer Vision, Proceedings of the International Congress of Mathematicians, ISBN 978-082-1801-10-9,Berkeley, CA, August, 1986

Hazel, G. G. (2000). Multivariate Gaussian MRF for Multispectral Scene Segmentation and Anomaly Detection. *IEEE Transactions on Geoscience and Remote Sensing*, Vol. 38, No. 3, (May 2000), pp. 1199–1211, ISSN 0196-2892

Huizenga, H., Munck, J., Waldorp, R., Grasman, R. (2002). Spatiotemporal EEG/MEG Source Analysis Based on a Parametric Noise Covariance Model. *IEEE Transactions on Biomedical Engineering*, Vol. 49, No. 6, (June 2002), pp. 533-539, ISSN 0018-9294

Jeng, F. and Woods, J. (1991). Compound Gauss-Markov Random Fields for Image Estimation. *IEEE Transactions on Signal Processing*, Vol. 39, No. 3, (March 1991), pp. 683–697, ISSN 1053-587X

Kashyap, R. and Chellappa, R. (1983). Estimation and Choice of Neighbors in Spatial-Interaction Models of Images. *IEEE Transactions on Information Theory*, Vol. 29, No. 1, (January 1983), pp. 60-72, ISSN 0018-9448

Kreyzig, E. (2005). *Advanced Engineering Mathematics, 8th. ed.*, Wiley, ISBN 978-047-1488-85-9, New York

Kyriakidis, P. C. & Journel, A. G. (1999). Geostatistical Space-Time Models: A Review. *Mathematical Geology*, Vol. 31, No. 6, (August 1999), pp. 651-684, ISSN 0882-8121

Li, S. Z. (1995) *Markov Random Field Modeling in Computer Vision*, Springer-Verlag, ISBN 978-4431701453, New York

Lu, N. & Zimmerman, D. (2005). The Likelihood Ratio Test for a Separable Covariance Matrix. *Statistics and Probability Letters*, Vol. 73, No. 4, (July 2005), pp. 449-457, ISSN 0167-7152

Magnus, J. R. & H. Neudecker (1999). *Matrix Differential Calculus with Applications in Statistics and Econometrics 2nd ed.*, Wiley, ISBN 978-047-1986-33-1, Chichester

Moura, J. M. F. & Balram N. (1993). Chapter 15: Statistical Algorithms for Noncausal Markov Random Fields, In: *Handbook of Statistics Volume 10*, Bose, N. K. & Rao, C. R., pp. 623-691, North Holland, ISBN 978-044-4892-05-8, Amsterdam

Naik, D. N. & Rao, S. S. (2001). Analysis of Multivariate Repeated Measures Data with a Kronecker Product Structured Covariance Matrix. *Journal of Applied Statistics*, Vol. 28, No. 1, (January 2001), pp. 91-105, ISSN 0013-1644;

Navarro, R. D. Jr., Magadia, J. C., & Paringit, E. C. (2009). Estimating the Gauss-Markov Random Field Parameters for Remote Sensing Image Textures, Proceedings of TENCON 2009 - 2009 IEEE Region 10 Conference, ISBN 978-142-4445-46-2, Singapore, November, 2009

Neudecker, H. (1969). Some Theorems on Matrix Differentiation with Special Reference to Kronecker Matrix Products. Journal of American Statistical Association, Vol. 64, No. 327, (September 1969), pp. 953-963, ISSN 0162-1459

Ravishanker, N. & Dey, D. K. (2002). *A First Course in Linear Model Theory*, CRC Press LLC, ISBN 978-158-4882-47-3, Boca Raton, FL

Richards, J. A. & Jia, X. (1999). *Remote Sensing Image Analysis: An Introduction, 4th ed.*, Springer-Verlag. ISBN 978-354-0251-28-6, Berlin

Rellier, G., Descombes, X., Falzon, F., & Zerubia, J. (2004). Texture Feature Analysis Using a Gauss–Markov Model in Hyperspectral Image Segmentation. *IEEE Transactions on Geoscience and Remote Sensing*, Vol. 42, No. 7, (July 2004), pp. 1543-1551, ISSN 0196-2892

Winkler, G. (2006). *Image Analysis, Random Fields and Dynamic Monte Carlo Methods: A Mathematical Introduction 2nd ed.*, Springer-Verlag. ISBN 978-354-0442-13-4, Berlin

6

Low Rate High Frequency Data Transmission from Very Remote Sensors

Pau Bergada, Rosa Ma Alsina-Pages, Carles Vilella
and Joan Ramon Regué
La Salle - Universitat Ramon Llull
Spain

1. Introduction

This chapter deals with the difficulties of transmitting data gathered from sensors placed in very remote areas where energy supplies are scarce. The data link is established by means of the ionosphere, a layer of the upper atmosphere that is ionized by solar radiation. Communications through the ionosphere have persisted, although the use of artificial repeaters, such as satellites, has provided more reliable communication. In spite of being random, noisy and susceptible to interference, ionospheric transmission still has favorable characteristics (e.g. low cost equipment, worldwide coverage, invulnerability, etc.) that appeal to current communications engineering.

The Research Group in Electromagnetism and Communications (GRECO) from La Salle - Universitat Ramon Llull (Spain) is investigating techniques for the improvement of remote sensing and skywave digital communications. The GRECO has focused its attention on the link between Antarctica and Spain. The main objectives of this study are: to implement a long-haul oblique ionospheric sounder and to transmit data from sensors located at the Spanish Antarctic Station (SAS) Juan Carlos I to Spain.

The SAS is located on Livingston Island (62.7 °S, 299.6 °E; geomagnetic latitude 52.6 °S) in the South Shetlands archipelago. Spanish research is focused on the study of the biological and geological environment, and also the physical geography. Many of the research activities undertaken at the SAS collect data on temperature, position, magnetic field, height, etc. which is temporarily stored in data loggers on-site. Part of this data is then transmitted to research laboratories in Spain. Even though the SAS is only manned during the austral summer, data collection never stops. While the station is left unmanned, the sets of data are stored in memory devices, and are not downloaded until the next Antarctic season. The information that has to be analyzed in almost real-time is transmitted to Spain through a satellite link. The skywave digital communication system, presented here, is intended to transmit the information from the Antarctic sensors as a backup, or even as an alternative to the satellite, without depending on other entities for support or funding.

Antarctica is a continent of great scientific interest in terms of remote sensing experiments related to physics and geology. Due to the peculiarities of Antarctica, some of these experiments cannot be conducted anywhere else on the Earth and this fact might oblige the researchers to transmit gathered data to laboratories placed on other continents for intensive study. Because of the remoteness of the transmitter placed at the SAS, the system suffers from power restrictions mainly during austral winter. Therefore, maintaining the radio link, even at a reduced throughput, is a challenge. One possible solution to increase data rate, with minimal power, is to improve the spectral efficiency of the physical layer of the radio link while maintaining acceptable performance. The outcomes and conclusions of this research work may be extrapolated to other environments where communication is scarcely possible due to economic or coverage problems. Therefore, the solutions presented in this study may be adopted in other situations, such as communications in developing countries or in any other remote area.

1.1 Remote sensors at the SAS

In this section we describe the main sensors located at the SAS, including a geomagnetic sensor, a vertical incidence ionosonde, an oblique incidence ionosonde and a Global Navigation Satellite System (GNSS) receiver. They have all been deployed in the premises of the SAS by engineers of the GRECO and scientist of the Observatori de l'Ebre. The geomagnetic sensor, the vertical incidence ionosonde and the GNSS receiver are commercial solutions. The oblique incidence ionosonde, used to sound the ionospheric channel between Antarctica and Spain, was developed by the GRECO in the framework of this research work.

1.1.1 Geomagnetic sensor

Ground-based geomagnetic observatories provide a time series of accurate measurements of the natural magnetic field vector in a particular location on the Earth's surface. This data is used for several scientific and practical purposes, including the synthesis and updates of global magnetic field models, the study of the solar-terrestrial relationships and the Earth's space environment, and support for other types of geophysical studies.

Once the raw observatory data is processed, it is sent to the World Data Centers, where the worldwide scientific community can access them. International Real-time Magnetic Observatory Network (INTERMAGNET) provides means to access the data by an almost real-time satellite link. The data is packed, sent to the geostationary satellites, and collected by Geomagnetic Information Nodes (GINs), where the information can be accessed freely. However, experience has shown that the satellite link is not 100% reliable, and it is preferable to have alternative means to retrieve the geomagnetic data.

There are three main reasons for designing a transmission backup system by skywave. Firstly, visibility problems appear when trying to reach geostationary satellites from polar latitudes. Secondly, end-to-end reliability can be increased by transmitting each frame repeatedly throughout the day. And finally, the ionospheric channel is freely accessed anywhere, whereas satellite communications have operational costs.

1.1.2 Ionosonde: vertical incidence soundings of the ionosphere

A vertical incidence ionospheric sounder (VIS) (Zuccheretti et al., 2003) was installed in order to have a sensor providing ionospheric monitoring in this remote region. This ionosonde is also being used to provide information for the High Frequency (HF) radio link employed for data transmission from the SAS to Spain. Data provided by the VIS is used to conduct ionospheric research, mainly to characterize the climatology of the ionospheric characteristics and to investigate the ionospheric effects caused during geomagnetically disturbed periods (see (Solé et al., 2006) and (Vilella et al., 2009)).

1.1.3 Oblique ionosonde

The oblique ionosonde monitors various parameters to model the HF radiolink between the SAS and Spain (Vilella et al., 2008). These parameters include link availability, power delay profile and frequency dispersion of the channel. The sounder includes a transmitter, placed on the premises of the SAS and a receiver deployed in Spain. The main drawback of the oblique sounder is the difficulty in establishing the ionospheric link. Firstly, the long distance of the link (12700 km) requires four hops to reach the receiver. And secondly, the transmitted signal has to cross the equator and four different time zones.

1.1.4 Global Navigation Satellite Systems

The ionosphere study can be approximated from several points of view. Vertical incidence soundings provide accurate information about electron density profiles below the peak electron density. However, when using this technique the electron profile must be extrapolated from the peak point to the upper limit of the ionosphere. Moreover, the low density of vertical ionosondes, especially in oceans and remote areas, is a serious impairment.

GNSS receivers constitute a high temporal and spatial resolution sounding network which despite gaps over oceans and remote regions, can be used to study fast perturbations affecting local regions, such as Travelling Ionospheric Disturbances and scintillations, or wider regions such as solar flares. Data gathered from GNSS receivers can provide information about the Total Electron Content (TEC) between receivers and satellites by means of proper tomographic modeling approaches. Spatial and temporal variations of the main ionospheric events can be monitored by means of GNSS receivers, especially those placed in the Antarctic Region, which is considered the entrance point of many ionospheric disturbances coming from Solar events. Furthermore, TEC reaches its highest variability peaks in the Antarctica area.

1.2 Data transmission

This chapter will study, analyze and experimentally verify a possible candidate for the physical layer of a long-haul ionospheric data link, focusing on the case SAS-Spain. Preliminary studies of data transmission feasibility over this link were already performed in (Deumal et al., 2006) and (Bergada et al., 2009), with encouraging results.

The first application of this link is the transmission of data generated by a geomagnetic sensor installed at the SAS. Future applications may include sending information of another nature such as temperature, glacier movements, seismic activity, etc.

The minimum requirements regarding the geomagnetic sensor data transmissions from the SAS to Spain are:

- The system should support a data throughput of 5120 bits per hour.
- The maximum delivery delay of the data should not exceed 24 hours.

1.2.1 Constraints

The extreme conditions prevailing at the SAS impose a number of restrictions that affect the transmission system. We highlight the following ones:

- The transmission power should be minimal. It is noted that the SAS is inhabited only during the austral summer, approximately from November to March. During this period there is no limitation regarding the maximum power consumption. However, the transmission system is designed to continue operating during the austral winter, when energy is obtained entirely from batteries powered by wind generators and solar panels. Hence the power amplifier is set to a maximum of only 250 watts.

- Environmental regulations applicable at the site advise against the installation of large structures that would be needed to install certain types of directive antennas.

1.2.2 Approach

This section justifies the need for a new data communication system adapted to the requirements of the project and presents the main ideas of this proposal. Firstly, we review the mechanisms that exist worldwide regarding the regulation of occupation of the radio spectrum. Then we review the features of current standards of HF data communications and discuss the non-suitability of these to the requirements of the project.

The International Telecommunication Union (ITU) is responsible for regulating the use of radio spectrum. From the point of view of frequency allocation, it has divided the world into three regions. Broadly speaking, region 1 comprises Europe and Africa, Asia and Oceania constitute region 2 and North and South America region 3.

In each region, the ITU recommends the allocation of each frequency band to one or several services. When multiple services are attributed to the same frequency band in the same region, these fall into two categories: primary or secondary. The ones that are classified as secondary services can not cause interference with the primary services and can not claim protection from interference from the primary services; however, they can demand protection from interference from other secondary services attributed afterwards.

Given these considerations, we propose a system transmission with the following guidelines:

- It can not cause harmful interference to any other service stations (primary or secondary).
- It can not claim protection from interference from other services.

To meet these requirements, we propose a system with the following characteristics:

- Reduced transmission power (accordingly with the consumption constrains).
- Low power spectral density.
- Robustness to interference.
- Burst transmissions (few seconds).
- Sporadic communications.

Moreover, given the ionospheric channel measures described in (Vilella et al., 2008) the following additional features are required:

- Robustness against noise (possibility of working with negative signal to noise ratio).
- Robustness against time and frequency dispersive channels.

1.2.3 HF communication standards

In this section we briefly review the current communication standards for HF and we justify its non-suitability for the purposes of this project.

Due to the proliferation of modems in the field of HF communications, interoperability between equipment from different manufacturers became a problem (NTIA, 1998). Hence the need to standardize communication protocols. Worldwide, there are three organizations proposing standards regarding HF communications: (*i*) the U.S. Department of Defense proposes the Military Standards (MIL-STD-188-110A, 1991; MIL-STD-188-110B, 2000; MIL-STD-188-141A, 1991), (*ii*) the Institute for Telecommunications Science (ITS), which depends on the U.S. Department of Commerce, writes the Federal Standard (FED-STD) and (*iii*) NATO proposes the Standardization Agreements (STANAG-4406, 1999; STANAG-5066, 2000).

Regarding the interests of this work it is noted that:

- The standard modes are designed for primary or secondary services. Therefore:
  - The bandwidth of the channels is standardized (3 kHz or multiples). Interference reduction, i.e. minimize the output power spectral density, with other transmitting systems is not considered.
  - No modes are considered based on short sporadic burst transfers to reduce interference with other users.
  - There are anti-jamming techniques (see MIL-STD-188-148) for additional application on a appropriate communication standard, but the proposals are not based on intrinsically robust to interference modulations.
- Robust configurations require a minimum signal to noise ratio (SNR) of 0 dB at 3 kHz bandwidth, which is not common in this link under the specified conditions of transmitted power and antennas (Vilella et al., 2008).

We conclude that the configurations proposed by current standards do not meet the desirable characteristics for the type of communication that is required in this work, and consequently, a new proposal should be suggested. In this chapter, we study a number of alternatives based on the use of Direct Sequence Spread Spectrum (DS-SS) techniques in order to cope with the impairments of the channel, the environment and other services.

2. Data transmission with Direct Sequence Spread Spectrum techniques

Spread Spectrum (SS) techniques are described by (Pickholtz et al., 1982) as a kind of transmission in which signal occupies a greater bandwidth than the necessary bandwidth to send the information; bandwidth spreading is achieved by an independent data source, and a synchronized code in the receiver to despread and retrieve data.

SS began to be developed especially for military purposes in the mid twentieth century and has continued in the forefront of research to present, which is, nowadays, a key point for the 3G mobile cellular systems (Third Generation Partnership Project, 1999) and wireless systems transmitting in free bands (IEEE802.11, 2007).

In the field of HF communications new techniques have always been slowly introduced due to a widespread sense that reliable communications were not feasible in this frequency band, while improvements of its implementation would be irrelevant. However, SS techniques have been suggested several times as suitable for the lower band of frequencies (i.e. LF, MF and, by extension, HF) (see (Enge & Sarwate, 1987)), since the intrinsic characteristics of SS systems to cope with multipath and interference (typical ionospheric channel characteristics).

There are three types of spread spectrum systems (Peterson et al., 1995): Direct Sequence, Frequency Hopping and hybrid systems composed by a mixture of both. In this study we will focus on Direct Sequence schemes.

DS-SS systems spread the spectrum by multiplying the information data by a spreading sequence. Consider the following model (Proakis, 1995):

$$s_{ss}(t) = \sum_{i=0}^{N_s-1} d_i c(t - iT_s), \ c(t) = \sum_{l=0}^{L-1} c_l p(t - lT_c), \tag{1}$$

where d_i denotes the i_{th} symbol, of length T_s, of a modulated signal:

$$\bar{d} = \{d_0, d_1, ..., d_{N_s-1}\} \tag{2}$$

and c_l are the chips [1], of length T_c, of a spreading sequence of length L:

$$\bar{c} = \{c_0, c_1, ..., c_{L-1}\} \tag{3}$$

and $p(t)$ is a pulse shaping defined as

$$p(t) = \begin{cases} 1, t \in [0, T_c) \\ 0 \Rightarrow \text{otherwise} \end{cases} \tag{4}$$

In addition, it holds that $LT_c = T_s$ and thus if the base band signal is formed by the symbols d_i and occupies a bandwidth of $\frac{1}{T_s}$ the spread spectrum signal $s_{ss}(t)$ occupies a bandwidth of $\frac{1}{T_c} = L\frac{1}{T_s}$.

The spreading sequence $\bar{c}$ should have good properties of autocorrelation and cross-correlation in order to ease the detection and synchronization at the receiver side.

Some of the main advantages of a system based on DS-SS are: (*i*) jamming and interference robustness, (*ii*) privacy, (*iii*) ability to use Code Division Multiple Access (CDMA) and (*iv*) robustness against multipath and time variant channels. On the other hand, the drawbacks of this technique are: (*i*) bandwidth inefficiency and (*ii*) receiver complexity: chip-level synchronization, symbol despreading (DS-SS signaling) and channel estimation and detection (RAKE receiver) (Viterbi, 1995).

Throughout the following sections we will discuss the most important considerations that justify the choice of DS-SS; as well as the technical basis to design the data modem for the ionospheric link between the SAS and Spain.

---

[1] The bits of a spreading sequence are called chips

2.1 Robustness against interference

Ionospheric communications have global coverage range. Consequently, any system operating in a given area might potentially interfere with other remote systems operating at the same frequency band. Hence the transmission system proposed in this work might be interfered with primary or secondary services that are assigned the same frequency band. For these reasons it is appropriate to review the characteristics of DS-SS regarding robustness against interference.

Let a DS-SS based system that transmits R_b bits per second in a bandwidth B_{ss} ($B_{ss} \gg R_b$) in the presence of additive white Gaussian noise $z(t)$ with power spectral density N_o $[W/Hz]$ and narrowband interference $i(t)$ with power P_i. At the receiver side:

$$r_{ss}(t) = s_{ss}(t) + i(t) + z(t). \tag{5}$$

Then (Pickholtz et al., 1982),

$$\left(\frac{E_b}{N_o}\right)_{z(t),i(t)} = \frac{P}{P_n}\frac{B_{ss}}{R_b}\frac{P_n}{P_n + P_i} = \frac{P}{P_n}\frac{B_{ss}}{R_b}\frac{N_0}{N_0 + \frac{P_i}{B_{ss}}}, \tag{6}$$

where $P_n = B_{ss}N_o$ is the noise power within the transmission bandwidth and $P = E_b R_b$ is the signal power. We can deduce from Equation 6 that we can reduce the effect of interfering signals by increasing B_{ss}. In other words, as $B_{ss} = L \cdot R_b$, the larger the spreading factor the lower the degradation due to interfering signals. The quotient $\frac{B_{ss}}{R_b}$ is called the process gain G_p and is a measure of the robustness of a spread spectrum system against interference. In DS-SS systems the processing gain coincides with the spreading sequence length (L).

It is noted that when B_{ss} increases $\left(\frac{E_b}{N_o}\right)\Big|_{z(t)}$ does not change because $P_n = N_0 B_{ss}$ increases in the same proportion, whereas an increase of B_{ss} implies an equivalent improvement of $\left(\frac{E_b}{N_o}\right)\Big|_{i(t)}$, as P_i is unchanged. To summarize, the use of spread spectrum provides improvement regarding narrowband interfering signals whereas no improvement over noise is achieved.

Feasibility studies of DS-SS systems with different types of interference can be found in the literature. See, for instance, (Schilling et al., 1980) when the interference is a narrowband signal and (Milstein, 1988) for multiple interfering signals.

2.2 Robustness against multipath channels

According to the analysis described in (Vilella et al., 2008), the ionospheric channel established between the SAS and Spain shows a maximum multipath delay spread (τ_{max}) that varies, depending on time and frequency, between 0.5 ms and 2.5 ms. Therefore, the coherence bandwidth of the channel, which can be considered as approximately the inverse of the maximum multipath delay spread (Proakis, 1995), can be narrower than 400 Hz. In case of transmitting with a wider bandwidth the channel would be frequency selective and distortion due to multipath would arise. Below, the properties of DS-SS against multipath are discussed.

Let a DS-SS based system with bandwidth B_{ss} in a channel with coherence bandwidth $W_c \sim \frac{1}{\tau_{max}} \ll B_{ss}$. If symbol time $T_s \gg \tau_{max}$ intersymbol interference due to multipath can be neglected. Moreover, if $T_s \ll \frac{1}{v_{max}}$ (where v_{max} denotes the maximum Doppler spread) the channel is almost invariant during a symbol time. Under these conditions it can be shown that (Proakis, 1995):

$$r_{ss}^{(k)}(t) = \sum_{n=1}^{N} h\left(\frac{n}{B_{ss}}\right) s_{ss}^{(k)}\left(t - \frac{n}{B_{ss}}\right) + z(t), \qquad (7)$$

where $h\left(\frac{n}{B_{ss}}\right)$ denotes a coefficient of the equivalent low-pass of the channel impulse response, $s_{ss}(t)$ corresponds to the base band spread signal defined in Equation 1, (k) denotes the contribution due to symbol k, $N = \tau_{max} B_{ss}$ is the number of non zero channel taps (since $s_{ss}(t)$ has a limited bandwidth of B_{ss}) and $z(t)$ is additive white Gaussian noise. In consequence, the signal reception is formed by delayed replicas of the transmitted signal. Then, we substitute Equation 1 in Equation 7 and apply an array of correlators to correlate the received signal with N copies of the spreading sequence $\bar{c}$ (each of them delayed a chip time). Let $\bar{c}$ be a sequence with good properties of circular autocorrelation:

$$\rho(m) = \sum_{l=1}^{L} c_l c_{l+m} \begin{cases} 1, & m = 0 \\ 0 \Rightarrow \text{otherwise} \end{cases} \qquad (8)$$

The output U_m of each correlator can be expressed as:

$$U_m = d_k h\left(\frac{m}{B_{ss}}\right) + \int_0^{T_s} c(t) z(t + \frac{m}{B_{ss}})\, dt, \ m \in [0, N-1]. \qquad (9)$$

Therefore, at the output of each correlator we obtain each transmitted symbol (d_k) multiplied by a channel coefficient $h\left(\frac{m}{B_{ss}}\right)$ plus a noise term. Hence, the use of DS-SS can take advantage of different replicas of the signal if correctly combined. The most general linear combination is the criterion of Maximal Ratio Combining that chooses the coefficients that maximize instantaneous SNR (Peterson et al., 1995). To properly apply this method it is mandatory to know the coefficients of the channel. Alternatively, the outputs of the correlators can be equally weighed (Equal Gain Combining), thus simplifying the receiver at the expense of worse performance.

2.3 Transmission with low spectral density power

One of the requirements of the proposed transmission system consists in causing minimal interference with primary and secondary services. For this purpose we propose alternatives to minimize power spectral density. We should note that as process gain increases, DS-SS techniques enable transmission with arbitrarily low power density. Suppose the transmission of a data stream $\bar{d}$ using a bandwidth B_d and power P. Then the average power spectral density is $D = \frac{P}{B_d} \left[\frac{W}{Hz}\right]$. Under the same conditions of power consider the transmission of the same data stream with DS-SS ($s_{ss}(t)$) by means of a spreading sequence $c(t)$ of length L. Then, the spectral occupancy of $s_{ss}(t)$ will be at least $L \cdot B_d$ and the average power spectral density will be $D_{ss} = \frac{P}{L \cdot B_d} \left[\frac{W}{Hz}\right]$.

Therefore, the use of DS-SS involves an average reduction of power spectral density by a factor equal to the process gain $G_p = L$. Then, the spectral occupancy proportionally increases; however, it is not an inconvenience in this case since there is no limitation in this regard.

2.4 Flexibility regarding spectral efficiency

The signal model expressed by Equation 1 is able to transmit $k = \log_2 K$ bits $(b_0^{(0)} \ldots b_{k-1}^{(0)})$ modulated in d_i during a period T_s (K is the number of possible modulation symbols in $\bar{d}$ and k is the corresponding number of bits per symbol).

The spectral efficiency ($C_{ss} = k/(T_s \cdot B_{ss})$), expressed in $[bits/s/Hz]$ and defined as the ratio between bit-rate and transmission bandwidth, is G_p times lower than the non spreading system.

There are several alternatives to increase spectral efficiency without decreasing process gain (and hence, robustness to interference) at the expense of increasing computational cost of the receiver. In the following sections we describe two of them: DS-SS M-ary signaling and quadriphase spreading. We briefly present the signal model, a study of the probability of error and we note the spectral efficiency of each of them.

2.4.1 DS-SS M-ary signaling

Let a set of M spreading sequences $Q = \left\{ \bar{c}^{(1)}, \bar{c}^{(2)}, \ldots, \bar{c}^{(M)} \right\}$ that satisfy a certain correlation relationship (orthogonal or nearly orthogonal according to Equation 8). Suppose that a certain sequence v from the previous set ($v \in [1, M]$) is transmitted depending on the value of $m = \log_2(M)$ bits of information. Then

$$s_{ss}(t) = \sum_{i=0}^{N_s-1} d_i \sum_{l=0}^{L-1} c_l^{(v)} p(t - iT_s - lT_C) = \sum_{i=0}^{N_s-1} d_i c^{(v)}(t). \tag{10}$$

This technique is called DS-SS M-ary signaling (see, for example, (Enge & Sarwate, 1987) for orthogonal sequences). On the receiver side, the optimum demodulator correlates the received signal with a replica of each of the M possible sequences belonging to the set Q. A noncoherent detector will make a decision based on the computation of the maximum likelihood of the M envelopes at the output of each correlator. The probability P_s of detecting an incorrect sequence in the presence of only additive white noise is given by (Proakis, 1995):

$$P_s = \sum_{p=1}^{M-1} (-1)^{p+1} \binom{M-1}{p} \frac{1}{p+1} e^{-\frac{p}{p+1}(m+k)\frac{E_b}{N_o}}. \tag{11}$$

The probability P_1 of making an error in the demodulation of coded bits transmitted in a certain sequence can be computed from the following expression (Proakis, 1995):

$$P_1 = \frac{2^{m-1}}{2^m - 1} P_s. \tag{12}$$

Once the sequence is detected we proceed to compute the probability P_2 of making an error in the demodulation of the coded bits in $\bar{d}$:

$$P_2 = \frac{1}{2} P_s + (1 - P_s) Q \left(\sqrt{\frac{2}{k} \left(\frac{E_b}{N_o} \right)'} \right), \tag{13}$$

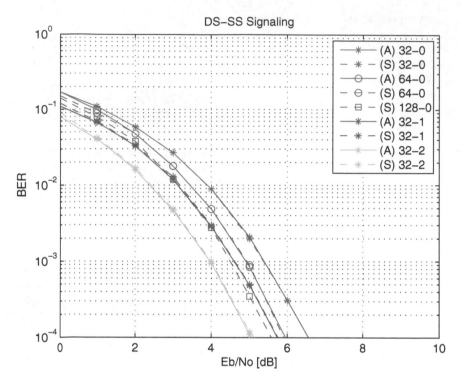

Fig. 1. Probability of error versus SNR per bit using DS-SS M-ary signaling for various values of M (32, 64, 128) and k ($k = 0$: no modulation, $k = 1$: BPSK, $k = 2$: QPSK). Probability is analytically (A) computed and derived from algorithm simulations (S)

where $\left(\frac{E_b}{N_o}\right)' = \left(\frac{E_b}{N_o}\right)(m + k)$. Finally, the joint probability P_b of bit error considering the contribution of both mechanisms is:

$$P_b = \frac{m \cdot P_1 + k \cdot P_2}{m + k}. \tag{14}$$

Figure 1 shows that the higher the M, the lower the SNR per bit required to obtain a certain BER. It can be explained by the fact that L increases as M (in a DS-SS system) and so does the process gain. It can be shown that the minimum SNR per bit required to obtain an arbitrarily small BER when $M \to \infty$ is -1.6 dB. Figure 1 also shows that the larger the k, the smaller the SNR per bit required to achieve a given BER. This apparent contradiction can be derived from the following two arguments: (*i*) for a given bit-rate, a high value of k enables the reduction of transmission bandwidth (and thus reduction of noise) and hence, improve the probability of finding the transmitted sequence (Equation 13). (*ii*) The probability P_b of total error (Equation 14) is a balance between P_1 and P_2. The second term in P_2 (Equation 13) derives from the probability of error in demodulating the bits in $\bar{d}$ once the sequence is successfully detected. So, if this term is lower than both the first term in Equation 13 and P_1 (Equation 12) the use of any kind of modulation will not result in significant degradation in P_b.

In a symbol time T_s we send $k + m$ bits $(b_0^{(1)} \cdots b_{k-1}^{(1)} b_0^{(1)} \cdots b_{m-1}^{(1)})$, where k corresponds to the bits modulated with $\bar{d}$ and m corresponds to the bits used in the choice of the spreading sequence. Consequently, the spectral efficiency is:

$$C = \frac{k + m}{G_p} = C_{ss} + \frac{m}{G_p} = C_{ss} + \frac{\log_2 M}{G_p}. \tag{15}$$

Therefore, the larger the M the lower the BER for a given SNR per bit and the greater the spectral efficiency at the expense of greater computational cost of the receiver. In addition, the larger the number of bits per symbol the better the BER for a given SNR per bit and the lower the computational cost of the receiver.

2.4.2 DS-SS M-ary signaling + Quadriphase

Another alternative is to divide the set of M sequences into two subsets: $Q_r = \{\bar{c}^{(1)}, \cdots, \bar{c}^{(M/2)}\}$ on one side and $Q_i = \{\bar{c}^{(M/2+1)}, \cdots, \bar{c}^{(M)}\}$ on the other side. Then, apply DS-SS M-ary signaling on both the real and the imaginary part of $\bar{d}$. Thus,

$$s_{ss}(t) = \sum_{i=0}^{N_s-1} \left(\Re\{d_i\} c^{(v_1)}(t) + j \cdot \Im\{d_i\} c^{(v_2)}(t) \right), \ v_1 \in [1, M/2], v_2 \in [M/2+1, M]. \tag{16}$$

This variant is called quadriphase chip spreading and permits us to send $M = 2 \log_2(M/2)$ bits per symbol by choosing a sequence from each of the two sets (plus k additional bits per symbol encoded in the modulation of d_i).

At the receiver side, the demodulator correlates the received signal with a replica of each of the M possible sequences. The detector will decide on the envelopes computed at the output of correlators corresponding to the sequences of the subset Q_r and a similar decision on the subset Q_i. The probability of incorrectly detecting a sequence from both the set Q_r and Q_i in the presence of only additive white Gaussian noise is:

$$P_s = \sum_{p=1}^{M/2-1} (-1)^{p+1} \binom{M/2-1}{p} \frac{1}{p+1} e^{-\frac{p}{p+1}\left(\frac{m+k}{2}\right)\frac{E_b}{N_o}}. \tag{17}$$

It is worth noting that the factor $1/2$ multiplying $\frac{E_b}{N_o}$ comes from considering that the symbol energy is equally distributed between real and imaginary parts (see Equation 16). The probability P_1 of incorrectly demodulating the bits coded in a sequence that belongs to the subset Q_r or Q_i can be obtained by applying an equation analogous to Equation 12:

$$P_1 = \frac{2^{(m/2)-1}}{2^{(m/2)} - 1} P_s. \tag{18}$$

We then discuss the probability P_2 of error on the bits modulated in $\bar{d}$ in the case of BPSK and QPSK. For BPSK, the decision is based on the sign of the sum of the two outputs of the correlators for the two sequences detected in the previous step. The case of QPSK is equivalent to two BPSK with half the SNR per each bit, both independently demodulated from the detection of two sequences (the corresponding to subsets Q_r and Q_i, respectively). Then, the probability P_2 for BPSK and QPSK, respectively, is:

$$P_2 = \frac{1}{2} P_s P_s + 2 P_s (1 - P_s) Q'_{bpsk} + (1 - P_s)(1 - P_s) Q_{bpsk}, \tag{19}$$

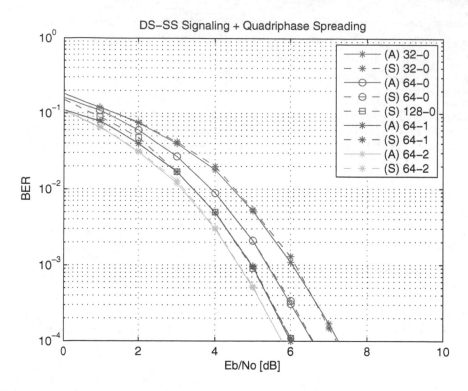

Fig. 2. Probability of error as a function of SNR per bit combining both techniques DS-SS M-ary signaling and quadriphase chip spreading for various values of M (32, 64, 128) and k ($k = 0$: no modulation, $k = 1$: BPSK, $k = 2$: QPSK). Probability is analytically (A) computed and derived from algorithm simulations (S)

$$P_2 = \frac{1}{2}P_s P_s + P_s(1 - P_s)\left(Q'_{bpsk} + 0.5\right) + (1 - P_s)(1 - P_s)Q'_{bpsk}, \tag{20}$$

where

$$Q_{bpsk} = Q\left(\sqrt{2\left(\frac{Eb}{N_o}\right)'}\right) \text{ and } Q'_{bpsk} = Q\left(\sqrt{\left(\frac{E_b}{N_o}\right)'}\right). \tag{21}$$

Finally, the probability P_b of bit error is equal to Equation 14. If we compare Figure 2 with Figure 1 it is shown that, for a given bit-rate, in terms of BER versus SNR per bit (for $k = 0$ with only additive white Gaussian noise) applying DS-SS M-ary signaling using M sequences is almost equivalent to using DS-SS M-ary signaling plus quadriphase chip spreading using $2M$ sequences. In this latter case, however, the process gain is doubled.

When we introduce modulation (i.e. $k \neq 0$) Figure 2 and Figure 1 show that the equivalence noted in the previous paragraph is no longer true: the use of quadriphase chip spreading with sequences of length $2M$ in combination with modulation produces inefficiency in terms of BER with respect to a system that does not use quadriphase chip spreading with sequences of length M. This is intuitively explained by noticing that when doubling the length of the sequence, keeping the same bandwidth, the number of transmitted sequences is halved as is the number of encoded bits in the modulation.

In a symbol time T_s, $k + m$ bits are sent $(b_0^{(1)} \cdots b_{k-1}^{(1)} b_0^{(1)} \cdots b_{m-1}^{(1)})$, k bits due to the modulation of $\bar{d}$ and m bits due the choice of the spreading sequence. Therefore, spectral efficiency is:

$$C = \frac{k+m}{G_p} = C_{ss} + \frac{m}{G_p} = C_{ss} + \frac{2 \log_2 (M/2)}{G_p} \tag{22}$$

Comparing Equation 22 with Equation 15 and equal bit rate, it is shown that quadriphase and biphase chip spreading have an approximate spectral efficiency (assuming $G_p = L \approx M$).

3. The experiments

This section describes the outcomes of various experiments based on DS-SS over the link established between the SAS and Spain. Firstly, we define the objectives of the study and point out some methodological criteria that was taken into account. Following, the testbench and the algorithms used to carry out the tests are described. Finally, the experiments are explained and the outcomes derived from them carefully discussed.

3.1 Goals

The aim of this work is to experimentally evaluate various alternatives, based on DS-SS, concerning the maximum achievable performance in terms of bit error rate and spectral efficiency at the expense of greater complexity at the receiver side. The final goal is to come up with a proposal for the data transmission link between the SAS and Spain. Therefore, the alternatives that we suggest may combine the following aspects:

- General features: (*i*) frequency chip, (*ii*) modulation.
- Related to DS-SS signaling: (*i*) process gain (determined by L), (*ii*) number of bits per sequence (expressed in terms of M), (*iii*) spreading: biphase or quadriphase.

However, there are a number of aspects, which are beyond the scope of this study, that must be defined and implemented. They are, specifically: (*i*) frame format, (*ii*) frequency and time synchronization (chip and frame), (*iii*) coding and interleaving and (*iv*) channel estimation and multipath diversity use.

It is noteworthy that it is not the aim of these experiments to measure the percentage of satisfactory receptions among the total number of receptions, since this magnitude is strongly related to the robustness of the synchronization method, which is beyond the scope of this study. Consequently, we will only evaluate expected performance from satisfactory receptions by means of a testbench explained below (see Section 3.3).

Figures 3 and 4 depict a block diagram of the transmitter and receiver, respectively. On one hand, common modules are shown in green. Specifically:

- At transmitter side: (*i*) a binary random source (320 bits), (*ii*) a turbo encoder (*rate* = 1/3) which operates combined with an interleaver (972 coded bits at the output), (*iii*) a frame compiler, designed according to the measured characteristics of multipath and Doppler spread, which builds a frame that consists of: (*iii.a*) a initial field for synchronization and channel estimation, (*iii.b*) a field that is periodically repeated to track channel estimation and (*iii.c*) data (see Section 3.3.1 and Figure 5).

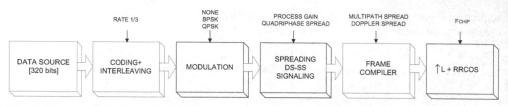

Fig. 3. Transmitter block diagram. Common modules to all experiments (testbench) are shown in green and modules with specific characteristics are shown in blue

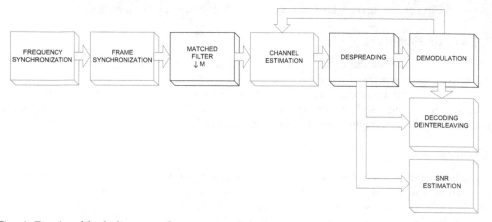

Fig. 4. Receiver block diagram. Common modules to all experiments (testbench) are shown in green and modules with specific characteristics are shown in blue

- At receiver side: (*i*) frequency synchronization by means of an unmodulated tone previously emitted, (*ii*) frame synchronization, (*iii*) channel estimation (*iv*) decoding and deinterleaving and (*v*) a SNR estimation module.

On the other hand, modules with specific parameters for the experiments are shown in blue in both the transmitter and the receiver. These parameters are: (*i*) chip frequency (f_{chip}), which determines the signal bandwidth (2500, 3125 and 6250 chips per second), (*ii*) modulation, a choice between no modulation, BPSK or QPSK, (*iii*) the process gain (spreading sequence of length 31, 63 or 127 chips), (*iv*) biphase or quadriphase spreading and (*v*) the number of bits per sequence $\log_2(M)$ (always $L = M - 1$).

3.2 Methodology

In this section we explain the approach followed prior to obtaining the outcomes from these experiments. We emphasize the following points:

- According to the explanations of the previous section, all experiments use a common testbench. Consequently, the test algorithms equally affect all experiments.
- Each experiment consists of a signal composed of 320 bits of data (972 coded bits), which are modulated, spread, filtered, and finally appropriate headers are appended to them. This signal has the appearance of a burst with a duration that depends on specific characteristics of the experiment (number of bits per symbol, sequence length,

etc.). Experiments are transmitted during a sounding period that has a maximum time length of 20 seconds, which is repeated every minute except for 18 minutes assigned to maintenance and other functions.

- In each sounding period several signals are transmitted within a frame. Each frame is repeated at least twice within a sounding period (more repetitions will be possible in case of short frames).

- Each sounding period is associated with a carrier frequency. Seven different carrier frequencies have been chosen based on availability outcomes presented in (Vilella et al., 2008). These carrier frequencies are: $\{8078, 8916, 10668, 11411, 12785, 14642, 16130\}$ $[kHz]$. Then, each frequency is tested 6 times per hour.

- Each day consists of 18 available hours (from 18 UTC to 11 UTC, both included).

- Each frame is transmitted a minimum of two days. Under these assumptions, each experiment was performed at a certain time and frequency, at least 24 times (2 days, 6 times per hour, 2 frames per sounding period).

- There are a number of days with frames containing a common experiment. This fact allows the assessment of interday variability.

3.3 Testbench

The testbench consists of a frame and a set of algorithms shared between all experiments, which are all described below.

3.3.1 Frame compilation

The testbench is based on a frame which is shown in Figure 5, where:

- C is a header based on two identical sequences $\bar{s}$ of length $L^{(s)}$ chips, as follows:

$$\mathbf{C} = \left\{ s_{L^{(s)}-l+1} \cdots s_{L^{(s)}} \, \bar{s} \, \bar{s} \, s_1 \cdots s_l \right\}. \tag{23}$$

Therefore, C has a length of $2L^{(s)} + 2l$ chips, where l is the number of chips circularly added before the first and after the second sequence. This header is used to achieve frame, chip and sample synchronization as well as initial channel estimation. The value of l can be computed by means of the maximum multipath spread of the channel (τ_{max}) as:

$$l = \lceil \tau_{max} f_{chip} \rceil, \tag{24}$$

where $\lceil \cdot \rceil$ denotes the integer immediately above. Therefore, l is the number of guard chips before and after the block formed by the two sequences $\bar{s}$. This guard ensures both circular correlation during synchronization and channel estimation free from intersymbol interference.

- S is a signaling field based on sequence $\bar{s}$, with the following form:

$$\mathbf{S} = \left\{ s_{L^{(s)}-l+1} \cdots s_{L^{(s)}} \, \bar{s} \, s_1 \cdots s_l \right\}. \tag{25}$$

Therefore, S is of length $L^{(s)} + 2l$. The value of l is calculated using Equation 24. This field provides channel estimation tracking. The period of repetition of S (denoted by T_S) is computed by means of the maximum Doppler spread of the channel (v_{max}) as:

$$T_S \approx \frac{1}{10 \, v_{max}}, \tag{26}$$

Fig. 5. Testbench frame format

where the channel is considered to be flat over a tenth of the inverse of v_{max}.

- **D** is a data symbol based on a Gold sequence of length L chips. Between the header **C** and the field **S**, or between two consecutive **S** fields there are B symbols that build a block. The number of symbols per block is given by the following equation:

$$B = round\left(\frac{T_S \cdot f_{chip}}{L^{(s)}}\right). \tag{27}$$

3.3.2 Algorithms description

This section explains reception algorithms used by all experiments (in green in block diagram of Figure 4).

Let $r[n]'$ be the signal at the output of a downsampling filter during the sounding period. $r[n]'$ is Δt seconds long with $\Delta t \cdot f_m$ samples, where f_m is the sampling frequency at the receiver side ($f_m = 50\,ksps$).

Estimation of frequency synchronization error (δf) between transmitter and receiver is obtained by applying algorithms explained in (Vilella et al., 2008) to a non modulated signal which is transmitted immediately before the data signal. Then, the signal $r[n]'$ is downconverted to baseband by a complex exponential signal with frequency $-\delta f$:

$$r[n] = r[n]' \cdot e^{2\pi\frac{\delta f}{f_m}n} \tag{28}$$

The next point to be considered is the frame, chip and sample synchronization which is obtained from the header **C** (Alsina et al., 2009). Firstly, emitter and receiver are time synchronized by means of a GPS receiver at each side, with time resolution of one second. Hence, the receiver knows the second t_a in which an experiment is transmitted. Let a synchronization window around t_a : $[t_a - \delta_a/2, t_a + \delta_a/2]$. Then the frame, chip and sample synchronization point t_s is:

$$t_s = \frac{\underset{m}{argmax}\left(\|S_1\| + \|S_2\|\right)}{f_m}, \ m \in [t_a - \delta_a/2, t_a + \delta_a/2]f_m, \tag{29}$$

where

$$S_1 = \sum_{k=0}^{L^{(s)}-1} r[m+k]\bar{s}[k] \text{ and } S_2 = \sum_{k=0}^{L^{(s)}-1} r[m + L^{(s)}\frac{f_m}{f_{chip}} + k]\bar{s}[k], \tag{30}$$

where $\bar{s}$ is the sequence of length $L^{(s)}$, interpolated by a root raised cosine filter, that forms header **C**.

It is noted that S_1 and S_2 are the correlation of the signal $\bar{r}$ with a replica of the header sequence $\bar{s}$ and with the same header sequence delayed $L^{(s)}$ chips, respectively. Therefore, synchronization probability is maximum for that value of m such that the sequences in S_1 and S_2 match in phase with header **C**.

It can be easily deduced from Equation 29 and Equations 30 that the greater the length of the sequence $\bar{s}$ (in chips and samples), the greater the likelihood of synchronization; however, the greater the length of the header. So, there is trade-off between synchronization performance and spectral efficiency.

Once frame, chip and sample (t_s) are successfully synchronized a matched filter is applied, followed by a decimate process to adjust the signal to one sample per chip (see Figure 4):

$$r_d[k] = \sum_{l=0}^{N_p-1} r[\frac{t_s}{f_m} + k\frac{f_m}{f_{chip}} + l]p[l], \tag{31}$$

where $\bar{p}$ is a pulse of length N_p samples, namely a root raised cosine with roll-off factor $\alpha = 0.65$.

Channel estimation is initally obtained from the second sequence on header **C** and tracked by the field **S** as:

$$h_l = \sum_{k=0}^{L^{(s)}-1} r_d[\frac{\delta t}{f_{chip}} + k + l]s[k], \; l \; in[-\tau_{max}f_{chip}, \tau_{max}f_{chip}], l \in Z, \tag{32}$$

where δt denotes the time offset of the sequence (**C** or **S**) from which we obtain channel estimation.

The despreading of each symbol is achieved by a bank of correlators using each of the sequences $\bar{c}^{(m)}$ that belongs to the family denoted by $Q = \{\bar{c}^{(1)}, \bar{c}^{(2)}, \cdots, \bar{c}^{(M)}\}$. The correlation is calculated for all those l values such that the channel estimation exceeds a certain threshold γ:

$$U^{(m)(l)} = \sum_{k=0}^{L-1} r_d[\frac{t_d}{f_m} + l + k]c^{(m)}[k], \; m \in [1, M], \; \forall l \mid \|h_l\| \geqslant \gamma, \tag{33}$$

where t_d indicates the starting point of the symbol under consideration and L is the length of the sequences used to spread data.

When using quadriphase spreading, the set of sequences Q is divided into two subsets $Q_r = \{\bar{c}^{(1)}, \cdots, \bar{c}^{(M/2)}\}$ on one side and $Q_i = \{\bar{c}^{(M/2+1)}, \cdots, \bar{c}^{(M)}\}$ on the other side. Then we compute both decision variables similarly to Equation 33.

The decision of which sequence has been transmitted is performed based on a criterion of maximum absolute value at the output of the correlators. It is only evaluated over the set or subset of appropriate sequences and for the shift l such that the channel estimation is maximum. We denote by p ($p \in [1, M]$) the index for the sequence with maximum correlator output, when not using quadriphase spreading, and p_r ($p_r \in [1, M/2]$) and p_i ($p_i \in [M/2+1, M]$) when using quadriphase spreading.

The demodulation of bits contained in $\bar{d}$ (see Equation 1) is achieved using a RAKE architecture. If not using quadriphase spreading, the decision is based on the decision variable U computed as follows:

$$U = \sum_l h_l^* \left(U^{(p)(l)} - \sum_{k<l} U^{(p)(k)} \rho^{(p)}(l-k) \right), \; \forall l \mid \|h_l\| \geqslant \gamma, \tag{34}$$

where $\rho^{(p)}$ is the circular autocorrelation of sequence p. If applying quadriphase spreading two decision variables (U_r and U_i) will be needed, one per each branch.

The value of p (or p_r and p_i) determines the bits used by the technique of spread spectrum, and the decision on U (or U_r and U_i) determines the bits used by modulation of $\bar{d}$.

If not using quadriphase spreading, each bit mapped to a symbol is linked to a soft-bit Sb that is computed according to the following expression:

$$Sb = \frac{\left\| U^{(p)(l)} \right\|^2}{\frac{1}{M-1} \sum\limits_{m=1,m\neq p}^{M} \left(\left\| U^{(m)(l)} \right| - \overline{U^{(l)}} \right)^2}, \; l \mid \forall k \neq l, \; \|h_l\| > \|h_k\|, \tag{35}$$

where:

$$\overline{U^{(l)}} = \frac{1}{M-1} \sum\limits_{m=1,m\neq p}^{M} \left\| U^{(m)(l)} \right\|. \tag{36}$$

It is noted that the term on the numerator in Equation 35 is a measure of the power of the signal after despreading, while the denominator is an estimation of the noise power, computed at the output of the correlators for those sequences which are not sent. Therefore, the soft-bit is an estimation of the signal to noise ratio after despreading. When using quadriphase spreading, soft-bits are calculated similarly to the biphase spreading option, for both detected sequences (p_r and p_i) and the corresponding subsets of sequences (Q_r and Q_i).

The noise variance is also computed at the output of the correlators except for those corresponding to the transmitted sequences. Once despreading and demodulation processes have finished (with the corresponding soft-bits) a deinterleaving and a Turbo decoding (Berrou & Glavieux, 1996) are applied. These two modules operate on a set of 972 coded bits and generate a set of 320 decoded bits. The Turbo code has a constraint length of 4 and runs 8 iterations.

If not using quadriphase spreading, SNR estimation is obtained averaging soft-bits values for each symbol of the burst. Specifically:

$$SNR = \frac{1}{N_{symbols}} \sum\limits_{n=0}^{N_{symbols}-1} \frac{Sb^{(n)}}{L}. \tag{37}$$

3.4 Outcomes

As a summary of the characteristics of most of the experiments carried out during the Antarctic season 2006/07 we have compiled Table 1. For each configuration we give the bandwidth (f_{chip}), the length of the sequence (L), the number of sequences (M), the use of quadriphase (QS), the type of modulation, the achieved bit rate, the spectral efficiency (C) (in parenthesis) and finally, the number of days each experiment was transmitted.

In order to summarize the outcomes obtained from the experiments carried out on the link between the SAS and Spain the plots shown in Figures 6 and 7 contain information from tens of thousands of bursts and are compared to the maximum achievable performance discussed in Section 2.4.

Config.	f_{chip}	L	M	QS	Modulation	bit rate (C)		Num. days
						uncoded	coded	
(1)	2500	63	64	0	none	238 (0.10)	79 (0.03)	1
(2)	2500	63	64	1	none	397 (0.16)	132 (0.05)	1
(3)	2500	63	64	1	QPSK	476 (0.19)	159 (0.06)	4
(4)	2500	31	32	1	QPSK	806 (0.32)	267 (0.11)	2
(5)	3125	63	64	0	none	298 (0.10)	99 (0.03)	1
(6)	3125	63	64	1	none	496 (0.16)	165 (0.05)	1
(7)	3125	63	64	1	QPSK	595 (0.19)	198 (0.06)	11
(8)	3125	31	32	1	QPSK	1008 (0.32)	336 (0.11)	2
(9)	6250	63	64	0	none	595 (0.10)	198 (0.03)	1
(10)	6250	63	64	1	none	992 (0.16)	331 (0.05)	1
(11)	6250	63	64	1	QPSK	1190 (0.19)	397 (0.06)	5

Table 1. Configurations of the experiments carried out on the ionospheric link between the SAS and Spain during the 2006/07 Antarctic season

The basic plot that is used to show the most important outcomes is a scatterplot (see, for instance, the two top pictures in Figure 6 containing $BER^{(')}$ performance versus SNR estimation. The estimation of SNR at the receiver side is computed immediately after despreading by means of Equation 37. Regarding this estimation it should be noted that (*i*) the signal strength is measured by means of only the most powerful path and hence, the signal at the receiver input is actually higher in case of multipath channel, (*ii*) when the detector at the output of correlators commits an error the subsequential SNR estimation is incorrect (see, for instance, Figure 6 (top) which shows that the detector systematically fails, producing $BER^{(')}$ close to 0.5 when the SNR is approximately -8 dB).

$BER^{(')}$ refers to the bit error rate measured on bits contained in a burst of N_{bits} (320 uncoded bits). Therefore, each point (SNR,$BER^{(')}$) of the scatterplot corresponds to the demodulation of a burst of N_{bits}. The thick line shown on each scatterplot is obtained by calculating the median of points of $BER^{(')}$ in consecutive subintervals of width 0.02.

The relationship between BER and SNR can be obtained by simulation, or analytically, according to the explanations in Section 2.4. Then, the probability P that a burst of N_{bits} contains k erroneous bits is:

$$P\left(BER^{(')} = \frac{k}{N_{bits}}\right) = \binom{N_{bits}}{k} BER^k (1 - BER)^{N_{bits}-k}. \tag{38}$$

We define the interval $\left[BER_l^{(')}, BER_h^{(')}\right]$ which, given a BER, contains with a probability of 90 % a defined $BER^{(')}$. Specifically:

$$P\left(BER^{(')} < BER_l^{(')}\right) = 0.05 \text{ and } P\left(BER^{(')} > BER_h^{(')}\right) = 0.05. \tag{39}$$

These scatterplots includes $BER_l^{(')} = f(SNR)$ and $BER_h^{(')} = f(SNR)$ curves for the analogous configuration. All the points should be found in 90 % of cases in the space between these curves if the tests were performed in a laboratory in the presence of only additive white Gaussian noise. However, as shown in Figures 6 and 7, it should be noted that in all scatterplots points are located outside the space bounded by the curves $BER_l^{(')}$ and $BER_h^{(')}$ and shifted about 2 dB to higher SNRs. This shift is due to different causes: (*i*) interference and no Gaussian noise, (*ii*) channel: multipath, Doppler, fading, etc., (*iii*) etc. The optimization of testbench algorithms could mitigate this loss of performance, but in any case we must take into account this shift when performing the design from a theoretical point of view.

Each scatterplot is accompanied by two histograms which derive from it. The first of these histograms shows, for each SNR, the percentage of receptions with $BER^{(')} = 0$ of the total number of receptions $BER^{(')} = 0$. It is noted that the higher the SNR the higher the probability of demodulating with $BER^{(')} = 0$, but simultaneously that SNR is less likely. This first histogram shows, therefore, the values of SNR at which the experiment is more successful. The second histogram shows, for each SNR, the percentage of receptions with $BER^{(')} = 0$ of the total number of receptions at that SNR. This figure allows us to evaluate at which SNR the probability of receiving a burst without errors is above a given value.

The results are discussed in terms of comparison with expected theoretical values. Specifically, in Figure 6 a scatterplot shows the effect of the variation in bandwidth and in Figure 7 the use of modulation is studied. Furthermore, in Figure 8 frequencies with best percentage of receptions of bursts with $BER^{(')} = 0$ per hour are shown and in Figure 9 the hours with best percentage of receptions of bursts with $BER^{(')} = 0$ at each frequency are also shown.

3.4.1 Bandwidth

Figure 6 compares the use of configuration (L, M, QS, Mod): (63, 64, yes, QPSK) with coded bits using a bandwidth of 3125 Hz (left column) and the same configuration using a bandwidth of 6250 Hz (right column). It is observed that the benefits obtained are slightly better for high bandwidth: for instance for SNR= -6 dB about 25 % of the receptions are $BER^{(')} = 0$ when $f_{chip} = 3125$ Hz, whereas this amount is over 40 % when $f_{chip} = 6250$ Hz (the percentages are also better in the second case for higher SNRs: -5 dB, -4 dB, -3 dB, etc.). This fact may be partly explained by a better performance of the RAKE receiver when working with higher multipath resolution.

3.4.2 Modulation

Figure 7 compares the application of QPSK modulation with a configuration with no modulation based on a system with (L, M, QS): (63, 64, yes) with coded bits and a bandwidth of 3125 Hz. Curves $BER_l^{(')}$ and $BER_h^{(')}$ indicate that theoretical maximum benefits are almost identical (slightly better when not using any modulation). The histograms confirm this estimation, where small deviations of about 5 % or 10 % to the no modulation option are observed.

It is worth noting that when using modulation the channel must be estimated and the use of a RAKE module is advised. Therefore, computational complexity is slightly increased while spectral efficiency improves without additional energy cost. In this context, we highlight the fact that the results of degradation of 2 dB observed between theory and experimental outcomes appear in both cases: modulation and no modulation. Therefore, this malfunction can be attributed to detection algorithms rather than the channel estimator and combiner algorithms.

3.4.3 Best frequencies

Figure 8 shows frequencies with best $BER^{(')}$ percentage, based on configuration (L, M, QS, Mod): (63, 64, yes, QPSK) with coded bits and a bandwidth of 3125 Hz. It should be noted that this configuration experimentally obtained $BER^{(')} = 0$ for SNR above -6 dB with probability greater than 80 % (see Figure 7). If we compare this figure with frequency availability results presented in (Vilella et al., 2008), which indicates the frequency with highest availability at a given SNR in a 3 kHz bandwidth, we can highlight that: (*a*) The distribution of frequencies with best availability rates is very similar in both studies: above 15 MHz between 18 and 22 UTC, from 9 MHz to 11 MHz between 23 and 6 UTC, and again about 15 MHz between 7 and 11 UTC. Therefore, there is a very good correspondence between channel sounding results and the analysis of data transmissions. (*b*) If we focus on specific values of percentages, we observe that (*b.i*) there are a set of hours, mostly belonging to the evening and morning (20, 21, 23, 2, 5, 6, 7, 8, 10 UTC) when the probability of overcoming -3 dB (measured by channel sounding) coincides, with high accuracy, with the probability of obtaining $BER^{(')} = 0$ (measured by data analysis). (*b.ii*) There are a number of hours at night (0, 1, 3, 4 UTC) when the probability of obtaining $BER^{(')} = 0$ is approximately 45 % below the prediction made by narrow-band sounding. (*b.iii*) Finally, a set of hours in both measures show mixed results (18, 22, 9 UTC). 18 and 9 UTC are noteworthy because the channel study shows very low availability (less than 5 %), whereas data analysis gets $BER^{(')} = 0$ with rates around 20%.

One possible explanation for these results could be derived from the following two arguments: (*i*) SNR measurements conducted by channel sounding consider noise everything that is not the transmitted signal (Gaussian noise and interference). During evening (18 to 22 UTC) and morning (07 to 11 UTC) the weight of interference power with respect to the total noise power is lower than during full night (23 to 06 UTC). It is precisely in the evening and morning when the two measurements (channel and data) are more similar. From this statement we can conclude that, rather than Gaussian noise, interference is the main factor on signal degradation. (*ii*) At full night and low frequencies (6 MHz to 10 MHz) channel time dispersion is greater than during evening and morning at high frequencies (14 MHz to 16 MHz) and, therefore, it is more difficult to obtain good performance for the same SNR (Vilella et al., 2008).

3.4.4 Best hours

Figure 9 shows the hours with highest percentage of $BER^{(')} = 0$ at each frequency, based on the configuration (L, M, QS, Mod): (63, 64, yes, QPSK) with coded bits and a bandwidth of 3125 Hz. This plot is especially useful when trying to use a directive antenna tuned to a particular frequency. It is found that the best results are achieved at high frequencies (around 16 MHz) in the early hours of night (21 UTC).

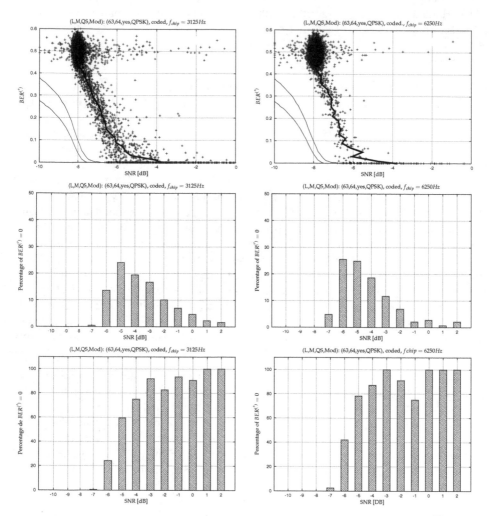

Fig. 6. Comparison of bandwidths (3125 Hz and 6250 Hz): (*i*) Scatterplot of $BER^{(')}$ versus SNR estimation before despreading (top row), (*ii*) histogram of the percentage of receptions with $BER^{(')} = 0$ to total receptions with $BER^{(')} = 0$ (middle row); (*iii*) histogram of the percentage of receptions with $BER^{(')} = 0$ to total receptions at that SNR (bottom row). The curves $BER_l^{(')}$ and $BER_h^{(')}$ are included on the scatterplots

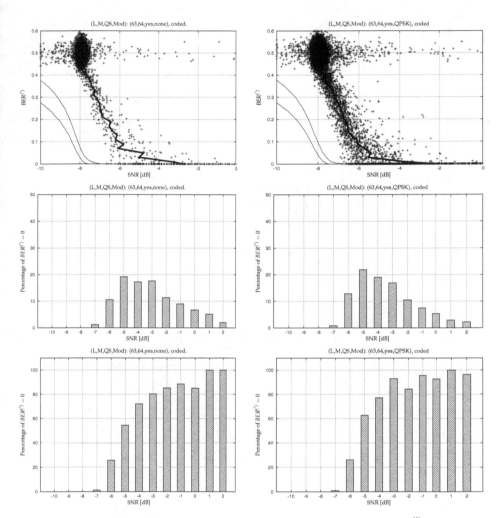

Fig. 7. Comparison of modulation (none and QPSK): (*i*) Scatterplot of $BER^{(')}$ versus SNR estimation before despreading (top row), (*ii*) histogram of the percentage of receptions with $BER^{(')} = 0$ to total receptions with $BER^{(')} = 0$ (middle row); (*iii*) histogram of the percentage of receptions with $BER^{(')} = 0$ to total measurements at that SNR (bottom row). The curves $BER_l^{(')}$ and $BER_h^{(')}$ are included on the scatterplots

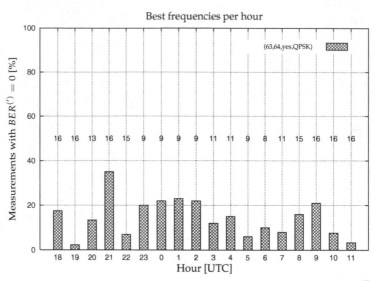

Fig. 8. Frequencies [MHz] with highest percentage of measurements with $BER^{(')} = 0$ per hour. The plot is based on the following configuration (L, M, QS, Mod): (63, 64, yes, QPSK) with channel coding

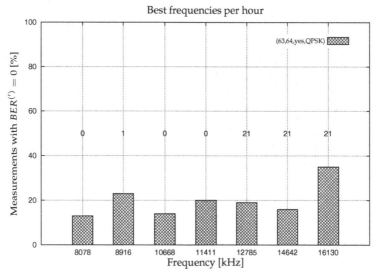

Fig. 9. Hours [UTC] with highest percentage of measurements with $BER^{(')} = 0$ at each carrier frequency. The plot is based on the following configuration (L, M, QS, Mod): (63, 64, yes, QPSK) with channel coding

4. Conclusions

Throughout this chapter we have studied, both theoretically and experimentally, the feasibility of low rate data transmission over a very long ionospheric link. The ionosphere may be used as a communications channel available from anywhere on the Earth. Hence it can be adopted as a solution to cope with deficient or non-existent satellite coverage range. We have focused our research work on the link between the Spanish Antarctic Base Juan Carlos I and Spain. It has a length of approximately 12700 km along the surface of the Earth and passes over 4 continents in a straight line. The system is currently applied to the transmission of data of a geomagnetic sensor that generates a maximum of 5120 bits per day. The special conditions found in Antarctica have impaired several aspects of the transmission. To conserve energy, maximum transmit power is set at 250 watts. In addition, to prevent further environmental impact, a non directive antenna (a monopole) requiring minimal infrastructure and installation was chosen to be placed at the SAS.

We have reviewed current HF communication standards and noted that none of them are intended for links with negative SNR. Thus we propose a novel system to be used on the physical layer of a ionospheric link based on a Direct Sequence Spread Spectrum technique. The determining factors for the use of this technique were its robustness to multipath and narrowband interference, its ability to transmit with low power spectral density, and its flexibility in terms of spectral efficiency in scenarios with negative SNR.

We propose a mode of transmission outside of current ITU standards, designed to cause minimal interference to primary and secondary services defined by the official agencies, able to operate in the presence of high values of noise power and interference, and robust to time and frequency channel dispersion. Hence, we suggest a transmission system based on sporadic short bursts of low density spectral power, focusing on increasing spectral efficiency and energy savings at the expense of a higher complexity receiver.

Several variants of DS-SS have been evaluated: signaling waveform, quadrature spreading and the impact of the modulation (BPSK and QPSK), all of them from the point of view of BER versus SNR per bit and spectral efficiency. We then conclude that:

- The DS-SS M-ary signaling technique allows an increase in spectral efficiency. The higher the number of sequences (M) the lower the SNR per bit required to achieve a given BER. In practice, if we use Gold spreading sequences, the maximum value of M is limited by the length of the spreading sequences ($M \sim L$). However, for a given bit-rate, if we increase M, the computational complexity at the receiver side increases.

- The combined use of modulation (BPSK and QPSK) and DS-SS M-ary signaling reduces the minimum required SNR per bit to achieve a certain BER. A greater reduction can be achieved with QPSK than with BPSK. However, modulation techniques require channel estimation (except for differential modulation) and, optionally, a RAKE combiner.

- If we add quadriphase spreading to DS-SS M-ary signaling (without modulation), gain can be doubled while maintaining BER and spectral efficiency performance. When using modulation, the use of quadriphase spreading results in energy inefficiency.

We assessed the suitability of studying a channel code based on the use of a Turbo code (rate = 1/3), with inner interleaver, that converts a burst of 320 bits into 972 coded bits. Simulations (not shown here for reasons of brevity) demonstrate that coding gain is only achieved for BER values below 10^{-4}. The reasons for using coding techniques will therefore depend, among other factors, on the size of the burst of bits and on the desired probability of error free.

We have defined a testbench to experimentally evaluate various configurations and to compare experiment outcomes with theoretical predictions. The testbench includes: (*i*) the definition of a header adapted to time and frequency channel dispersion to perform synchronization and channel estimation, (*ii*) the definition of a data frame, (*iii*) the design of a set of algorithms: encoding/decoding, synchronization, spreading/despreading, RAKE combiner, demodulator and SNR estimation.

The outcomes gathered from this testbench have shown that, for instance, with a SNR of -5 dB, this ionospheric data transmitter is able to transmit data (6 kHz and 320 bits burst size) with a rate of 397 bits per second (error free) with a successful probability of approximately 95 % (see Table 1 and Figure 6). It is noted that this rate would suffice to send the amount of data required by the application (5120 bits per hour), with sporadic frequency transmissions.

Experimental tests have been performed for different configurations and at different bandwidths in a frequency range between 8 MHz and 16 MHz and a time interval between 18 and 12 UTC. From the experimental results and comparison, with theoretical predictions in terms of BER versus SNR, the following conclusions can be drawn:

- There is a loss of about 2 dB of SNR between the theoretical and experimental BER. This loss may be attributable to several factors: non-Gaussian noise, interference, channel dispersion, and so on.
- For a given SNR, the probability of receiving a burst without error is slightly higher for higher bandwidths. This improvement may be due to better performance of the RAKE combiner due to higher multipath resolution (this result should be confirmed in later experiments).
- Experimental results confirm that for a given SNR at the receiver, the use of modulation added to signaling techniques (thus increasing the bitrate without increasing the transmitted power) does not affect the BER performance.
- Regarding the frequencies that are more likely to transmit error free bursts, we observe that they correspond with great accuracy to those with highest availability, measured by channel studies ((Vilella et al., 2008)): above 15 MHz in the evening (18 to 22 UTC) and morning (7 to 11 UTC), and below 11 MHz in the early morning (23 to 6 UTC). Regarding specific percentages of bursts without errors, it appears that they are very similar to those equivalent measurements done by channel studies during the evening and morning, but are worse at night and early morning. This is mainly attributed to the increased amount of interference at night.

According to experimental results we make the following recommendations: (*i*) integrate the loss of 2 dB of SNR into theoretical calculations, (*ii*) prioritize larger bandwidths, use modulation (QPSK rather than BPSK) and use coding techniques, (*iii*) use modulation plus M-ary signaling without quadriphase spreading, (*iv*) optimally attempt to establish the data link at 21 UTC (at 16 MHz), or from 23 to 6 UTC (within the range 9-11 MHz).

5. Acknowledgments

This work has been funded by the Spanish Government under the projects REN2003-08376-C02-02, CGL2006-12437-C02-01/ANT, CTM2008-03236-E/ANT,

CTM2009-13843-C02-02 and CTM2010-21312-C03-03. La Salle thanks the *Comissionat per a Universitats i Recerca del DIUE de la Generalitat de Catalunya* for their support under the grant 2009SGR459. We must also acknowledge the support of the scientists of the Observatory de l'Ebre throughout the research work.

6. References

Alsina, R. M., Bergada, P., Socoró, J. C. & Deumal, M. (2009). Multiresolutive Acquisition Technique for DS-SS Long-Haul HF Data Link, *Proceedings of the 11th Conference on Ionospheric Radio Systems and Techniques*, IET, Edimburgh, United Kingdom.

Bergada, P., Deumal, M., Vilella, C., Regué, J. R., Altadill, D. & Marsal, S. (2009). Remote Sensing and Skywave Digital Communication from Antarctica, *Sensors* 9(12): 10136–10157.

Berrou, C. & Glavieux, A. (1996). Near optimum error correcting coding and decoding: Turbo-codes, *IEEE Transactions on Communications* 44(10): 1261–1271.

Deumal, M., Vilella, C., Socoró, J. C., Alsina, R. M. & Pijoan, J. L. (2006). A DS-SS Signaling Base System Proposal for Low SNR HF Digital Communications, *Proceedings of the 10th Conference on Ionospheric Radio Systems and Techniques*, IET, London, United Kingdom.

Enge, P. K. & Sarwate, D. V. (1987). Spread-spectrum multiple-access performance of orthogonal codes: Linear receivers, *IEEE Transactions on Communications* 35(12): 1309–1319.

IEEE802.11 (2007). *Wireless LAN Medium Access Control (MAC) and Physical Layer (PHY) - Specifications (2007 Revision)*, number doi:10.1109/IEEESTD.2007.373646.

MIL-STD-188-110A (1991). *Interoperability and Performance Standards for Data Modems*, U.S. Department of Defense.

MIL-STD-188-110B (2000). *Interoperability and Performance Standards for Data Modems*, U.S. Department of Defense.

MIL-STD-188-141A (1991). *Interoperability and Performance Standards for Medium and High Frequency Radio Equipment*, U.S. Department of Defense.

Milstein, L. B. (1988). Interference rejection techniques in spread spectrum communications, *IEEE Transactions on Communications* 76(6): 657–671.

NTIA (1998). High frequency radio automatic link establishment (ALE) application handbook, *NTIA handbook*.

Peterson, R. L., Ziemer, R. E. & Borth, D. E. (1995). *Introduction to Spread Spectrum Communications*, Prentice Hall.

Pickholtz, R. L., Schilling, D. L. & Milstein, L. B. (1982). Theory of spread-spectrum communications - a tutorial, *IEEE Transactions on Communications* 30(5): 855–884.

Proakis, J. G. (1995). *Digital Communications*, McGraw-Hill.

Schilling, D. L., Milstein, L. B., Pickholtz, R. L. & Brown, R. W. (1980). Optimization of the processing gain of an M-ary direct sequence spread spectrum communication system, *IEEE Transactions on Communications* 28(8): 1389–1398.

Solé, J. G., Alberca, L. F. & Altadill, D. (2006). Ionospheric Station at the Spanish Antarctic Base: Preliminary Results (in Spanish), *Proceedings of the 5th Asamblea Hispano-Portuguesa de Geodesia y Geofísica*, Sevilla, Spain.

STANAG-4406 (1999). *Military Message Handling System (MMHS)*, North Atlantic Treaty Organization.

STANAG-5066 (2000). *Profile for High Frequency (HF) Radio Data Communications*, North Atlantic Treaty Organization.

Third Generation Partnership Project (1999). *Physical layer - General description Release'99*, number 3GPP TS 25.201, Technical Specification Group Radio Access Network.

Vilella, C., Miralles, D., Altadill, D., Costa, F., Solé, J. G., Torta, J. M. & Pijoan, J. L. (2009). Vertical and Oblique Ionospheric Soundings over a Very Long Multihop HF Radio Link from Polar to Midlatitudes: Results and Relationships, *Radio Sci.* 44(doi:10.1029/2008RS004001).

Vilella, C., Miralles, D. & Pijoan, J. L. (2008). An Antarctica-to-Spain HF ionospheric radio link: Sounding results, *Radio Sci.* 43(doi:10.1029/2007RS003812).

Viterbi, A. J. (1995). *CDMA: Principles of Spread Spectrum Communication*, Prentice Hall PTR.

Zuccheretti, E., Tutone, G., Sciacca, U., Bianchi, C. & Arokiasamy, B. (2003). Vertical and oblique ionospheric soundings over a very long multihop hf radio link from polar to midlatitudes: Results and relationships, *Ann. Geophys* (46): 647–659.

7

A Contribution to the Reduction of Radiometric Miscalibration of Pushbroom Sensors

Christian Rogaß[*] et al.[**]

Helmholtz Centre Potsdam, GFZ German Research Centre for Geosciences,
Germany

1. Introduction

Imaging spectroscopy is used for a variety of applications such as the identification of surface cover materials and its spatiotemporal monitoring. Contrary to multispectral instruments more spectral information can be incorporated in the differentiation of materials. New generations of sensors are based on the pushbroom technology, where a linear array of sensors perpendicular to the flight direction scans the full width of the collected data in parallel as the platform moves. Contrary to whiskbroom scanners that collect data one pixel at a time pushbroom systems can simply gather more light as they sense a particular area for a longer time. This leads to a better Signal-to-Noise Ratio (SNR). In addition, the two dimensional photo detector array in pushbroom systems may enable different readout configuration settings, such as spatial and/or spectral binning, allowing a better control of the SNR. It follows from this that low reflective materials can be potentially sensed as well as high reflective materials without saturating the detector elements. However, the use of detector arrays requires a precise radiometric calibration as different detectors might have different physical characteristics. Any miscalibration results in visually perceptible striping and uncertainties increase in preceding analyses such as classification and segmentation (Datt et al., 2003). There are various reasons for miscalibration, for instance temporal fluctuations of the sensor temperature, deprecated calibration coefficients or uncertainties in the modelling of the calibration coefficients. In addition, ageing and environmental stresses highly affect the mechanical and optical components of a sensor system; its reliability is thus not such to grant unchanged calibration accuracies for the entire mission life span.

Radiometric calibration and the estimation of the calibration coefficients can be considered as the assignment of known incident at-sensor radiance to measured digital numbers (DN). For this, physically known, different reflective targets are artificially illuminated by electromagnetic radiation of a specific spectrum and the reflected radiation is then recorded by the sensor that consists of a number of detectors. Then, the response of each detector is

---

[*] Corresponding Author
[**] Daniel Spengler[1], Mathias Bochow[1], Karl Segl[1], Angela Lausch[2], Daniel Doktor[2], Sigrid Roessner[1], Robert Behling[1], Hans-Ulrich Wetzel[1], Katia Urata[1], Andreas Hueni[3] and Hermann Kaufmann[1]
[1]*Helmholtz Centre Potsdam, GFZ German Research Centre for Geosciences, Germany*
[2]*Helmholtz Centre for Environmental Research, UFZ Germany*
[3]*Remote Sensing Laboratories, University of Zurich, Switzerland*

modelled with respect to the incident radiation, the reflective target and the defined illumination of the target. The mathematical modelling is often performed by applying a linear least squares regression. Contemporary, differences of detectors are balanced.

Consequently, calibration coefficients are obtained – shortly named as offset and slope. Offsets incorporate the unwanted detector-dependent dark current that is caused by thermally generated electrons (Oppelt and Mauser, 2007). In turn, slopes directly relate radiance to DN. Offsets are often measured before any image acquisition, but may change due to instabilities in the cooling system. Mechanical stress or uncertainties in foregoing laboratory calibration can cause changes in the physical characteristics of detectors as well. In order to support laboratory calibration, in-flight calibrations complement the calibration procedure, verifying the results obtained in the laboratory and, in addition, allowing the measurement of parameters that are only obtainable during flight (i.e. stability measurements, solar calibration, etc).

For this, physically known targets have to be sensed and incident illumination should be measured during the overflight. Uncertainties in the measurement of hemispheric incident solar radiation and in the incorporation of illumination, sensing and wavelength dependent response of imaged calibrations targets on incident light aggravate then this type of calibration and may also lead to miscalibrations or visually perceptible image stripes. Hence, any striping reduction or retrieval of calibration coefficients should reduce stripes and at the same time the spectral characteristics of the imaged surface materials have to be preserved.

In the literature, specific approaches for destriping of slope stripes, offset stripes or both exist, and these are primarily based on methods such as interpolation (Oliveira and Gomes, 2010; Tsai and Chen, 2008), local or global image moments (Datt et al., 2003; Cavalli et al., 2008; Le Maire et al., 2008; Liu et al., 2009), filtering (Garcia and Moreno, 2004; Shen et al., 2008; Simpson et al., 1995, Simpson et al., 1998) or complex image statistics of log transformed slopes (Bouali and Ladjal, 2010; Carfantan and Idier, 2010; Gomez-Chova et al., 2008). Most methods replace original, miscalibrated radiances. This should be only applied if information is completely missing or erroneous.

In the following, a framework that efficiently reduces linear as well as nonlinear miscalibration is reviewed concurrently preserving the spectral characteristics of sensed surface cover materials. This framework, originally proposed by Rogass et al. (2011) and named as Reduction of Miscalibration Effects (ROME), consists of a linear and a nonlinear slope reduction and an offset reduction that are consecutively performed and does not require a priori information or scene and sensor specific parameterisation.

Before any radiometric miscalibration reduction is applied, image gradients that are not orthogonal to the image are excluded if they do not represent the image content. Here, Minkowski metrics, gradient operators and edge extraction algorithms are combined to exclude discontinuities if they do not dominate the image content (Canny, 1986; Haralick et al., 1987; Rogass et al., 2009). The linear and the nonlinear slope reduction of ROME are performed for each detector element and band without any information from other detector elements. The offset reduction of ROME considers adjacent image columns and refers to a predefined image column (first column per default) that is assumed to be the reference. Specific image quality metrics, such as the change in SNR (Gao, 1993; Atkinson et al., 2005), were used to evaluate the necessity of such preceding reduction.

After these preceding reductions the image is radiometrically band wise rescaled to recover the radiometric scale. This is necessary since uncertainties in the estimation of parameters (e.g., detector resolution in the linear slope reduction) and in the incorporation of miscalibrated reference areas (e.g., potential miscalibration of the first image column as reference for the offset reduction) remain. The rescaling of ROME assumes that image columns that were less corrected than others can be used as reference for the whole image. After all reductions a detrending is performed reducing across track brightness gradients caused by reduction related frequency undershoots of low SNR bands. In this work an extension of ROME's detrend approach is presented evidencing an effective reduction of undershoots when compared to the original approach.

In order to test the robustness of the algorithm due to different types of miscalibration, four grey valued images as well as 12 multispectral and hyperspectral scenes were considered. The grey valued images were randomly striped by linearly varying slope and/or offset. One HyMAP scene was three times differently and artificially striped by offset stripes. The simulated EnMAP scene was not corrected for nonlinear effects and, hence, the nonlinear correction facilities were tested. Miscalibrated scenes acquired by AISA DUAL (3 scenes), Hyperion (2 scenes), ASTER (1 scene), CHRIS/Proba (1 scene) and APEX (1 scene) were additionally processed.

2. Materials

In Rogass et al. (2011) four grey valued images (Fig. 1) from the image database of the Signal and Image Processing Institute (SIPI) of the University of California (Weber, 1997), 512 × 512 pixels in size, and six hyperspectral scenes (3 AISA DUAL, 2 Hyperion and 1 EnMAP) were selected to test and to evaluate the performance of the proposed ROME framework. The grey valued samples as well as the EnMAP scene were considered as noise free. However, the 'Lenna' image (Fig. 1a) and the 'Mandrill' image (Fig. 1b) are excluded from further considerations due to their unique spectral and spatial properties as detailed described in Rogass et al. (2011).

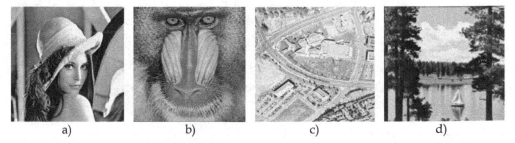

a)   b)   c)   d)

Fig. 1. Grey scaled image samples from the USC SIPI image data base considered in the following as a) 'Lenna', b) 'Mandrill', c) 'Aerial' and d) 'Sailboat on lake'

To simulate different types of miscalibrations and to evaluate their impact on the proposed work, the two grey valued images (Fig. 1 c and d) and the EnMAP scene were artificially degraded. The grey valued images were randomly degraded by applying 800 different sets of multiplicative (slope) and/or additive (offset) Gaussian white noise (Box and Muller,

1958). These 800 noisy matrices were transformed to provide always a mean equal to zero and standard deviations ranging from 0.0001 to 10000 for the multiplicative parts and from -10000 to 10000 for the additive part. Such high noise levels were chosen to also simulate low SNR scenarious that are noise dominated. More details on the noise matrices and the hyperspectral scenes are given in Rogass et al. (2011).

In this work additional scenes from APEX, ASTER and CHRIS/Proba were inspected, destriped and evaluated. Contemporary, one HyMAP scene was selected and three times artificially and additively degraded by Gaussian white noise to extend the testing of correction facilities for airborne sensors. After degrading three mean SNR levels of 7.6, 76 and 760 were simulated.

The HyMAP sensor is a hyperspectral whiskbroom airborne sensor that consists of one detector column and, hence, offset miscalibrations cannot be perceived as image stripes since each image column has the same offset. Therefore, HyMAP image acquisitions can be used to test correction approaches for pushbroom sensors.

In the following, an image column or across track is considered as x and an image row or along track is considered as y.

3. Methods

3.1 Calibration basics

Radiometric calibrations are often performed in laboratory and basically assign known incident at-sensor radiance to measured digital number (DN). The association is usually realised by a linear least squares regression that minimises the difference between modelled at-sensor radiance and known at-sensor radiance. The regression coefficients are also used in the reverse process to assign measured DN to at-sensor radiance that is considered as radiometric scaling (Chander et al., 2009).

However, uncertainties in the laboratory measurements, in the mathematical modelling and in the incorporation of temporal changes of the detector characteristics lead to miscalibrations and, hence, to visually perceptible image stripes in y-direction. In the following it will be exemplarily shown how to suppress miscalibrations in accordance with the ROME framework. This framework consists of multiple steps that are consecutively processed (Fig. 2).

Fig. 2. Workflow of ROME destriping per band

Pushbroom sensors have detector arrays. Each detector pixel of the array has different physical characteristics. It follows from this that an uncalibrated hyperspectral image is striped. The radiometric calibration and the reverse process - radiometric scaling - aim at the assignment of incident radiance to DN and vice versa. Usually, radiometric calibration can be performed in-flight, vicariously (Biggar et al., 2003; Bruegge et al., 2007), over a flat field (Bindschadler and Choi, 2003) or in laboratory.

In the process of calibration each detector of the detector array must be solely considered. Known incident radiation reaches a detector pixel and once the incident photons have sufficient energy to excite electrons into a certain energy level, electron-hole pairs are generated – a phenomenon that is known as the photoelectric effect. These free charges are then transmitted and read out through sensor electronic. Dispersive optics placed in front of the sensor disperses the incident radiation into different wavelenghts that is further projected into each row of the detector array. The physical response, considered as signal S in electrons, of one detector element of a pushbroom sensor to incident radiation L can be approximated by a nonlinear relation (Dell`Endice, 2008; Dell`Endice et al., 2009):

$$S(e^-) \propto \frac{F \cdot L \cdot A \cdot \tan^2\left(\frac{FOV}{2}\right) \cdot \tau \cdot T \cdot \lambda \cdot \eta \cdot SSI}{h \cdot c \cdot n_{e^-}^2} \tag{1}$$

where L is the at-sensor-radiance, A is the optical aperture of the sensing instrument, FOV is the field of view, T is the integration time, SSI is the Spectral Sampling Interval in respect to the Full Width at Half Maxima, h is the Planck constant, c is the speed of light, n_{e^-} is the number of collected electrons, τ is the optical transmission, λ is the centre wavelength, η is the quantum efficiency and F is the filter efficiency. This can be then related to the recorded digital number DN as follows:

$$DN = \frac{(S+N) \cdot DN_{max}}{FWC} + DN_0 \quad \wedge \quad S \leq FWC \tag{2}$$

where N is a noise term incorporating Shot-Noise, read-out noise and dark noise, DN_{max} is the radiometric resolution, FWC is the Full Well Capacity that defines the detector saturation and DN_0 is the dark current. To enable a mathematical modelling relating incident radiation and measured DN, either the illumination is changed in a defined way or the integration time is changed or targets of different reflective properties are sensed. The association of at-sensor radiance L to DN is broadly considered as radiometric calibration or, reversely, as radiometric scaling (Chander et al., 2009). To reduce the influence of noise, a specific number of measurements is required. Then, the association can be realised, e.g., by least squares polynomial fit that minimises the differences between modelled and measured at-sensor radiance (Barducci et al., 2004; Xiong and Barnes, 2006). The minimisation of the merit function gives then the transformation coefficients for the association. This can be achieved by applying the following model:

$$\chi^2 = \sum_{j=1}^{N_{targets}} \left[L - \left(c_0 + \sum_{i=1}^{M} c_i \cdot DN^i \right) \right]^2 \quad \wedge \quad M \geq 1; \ N_{targets} \geq 2 \tag{3}$$

where $N_{targets}$ denotes the number of calibration targets, c_0 is the offset regarding the dark current, and M is the polynomial degree. The more the detector response differ from a linear response, the more it is necessary to use a polynomial degree higher as one. Mostly, detector responses can be mathematically modelled. Potential changes in the characteristics of detectors require frequent calibrations that are not practicable.

However, if then along track stripes in radiometrically scaled images are perceptible miscalibration is indicated. In that case, it is necessary to determine the type of miscalibration – multiplicative or additive – linear or nonlinear. In ROME this is performed by comparing the output SNR to the input SNR due to the specific processing step (Brunn et al., 2003; Gao, 1993). If the SNR is increased, a successful operation is indicated and finally applied. In the following the stripe types are distinguished with respect to equation 3 – additive c_0 and multiplicative $c_{1..M}$ miscalibration and reduction. In any case the reduction of miscalibration should be applied before rectification.

3.2 Edge exclusion

Discontinuities such as impulse noise, edges or translucent objects like tree vegetation should be excluded from further processing unless they contribute a high spatial distribution. This is relevant for approaches that aim on the reduction of miscalibration by relying on statistical analyses of spatial and spectral differences in homogeneous regions. Edges can be generally excluded if they do not coincide with along track or across track direction. Since uncertainties in the impact of edges on the reduction process remain edges should be excluded if they do not dominate image content (compare Fig. 1b). In ROME this is performed by a combination of edge detection algorithms with morphological dilation with respect to Minkowski metrics. Potential edge detection algorithms for single banded images must be then adapted to incorporate only along track gradients, because gradients of radiometric miscalibration might superimpose across track gradients. In Rogass et al. (2011) the Canny algorithm (Canny, 1986) is used for single banded images and the Hyperspectral Edge Detection Algorithm (HEDA) is used for multi banded images (Rogass et al., 2010). After obtaining binary edge maps morphological dilations (Haralick et al., 1987; Rogass et al., 2009) are additionally applied to minimise edge adjacency effects caused by Point Spread Function (PSF) related blooming of edges into adjacent regions. The reversed edge map gives then the mask. In case of tree vegetation indices are computed and pixel wise thresholded by the highest two likelihood quartiles of containing vegetation. This binary vegetation map is reversed and multiplied with the reversed binary edge map. Hence, edges and translucent vegetation is excluded. Related equations are given in Rogass et al. (2010) and Rogass et al. (2011). The application of the reverse edge map gives than an edge filtered image.

3.3 Linear c_1 slope reduction

In case of linear miscalibration each pixel of one detector (one column) of the same channel is scaled by the same c_1 slope (the term 'gain' is often misleading used and corresponds to the maximisation of the radiometric resolution; Chander et al., 2009). A simple differential operation between two pixels from the same column leads to the mathematical elimination of the c_0 offset. This difference is then equivalent to the difference of radiance levels. This corresponds to the c_1 slope of this detector times the spectral difference of surface cover materials constrained by the detector resolution. Hence, a reduction of c_1 miscalibration must recover both c_1 slope and the spectral characteristics of the surface cover material. In ROME this is performed per detector or column and band by applying a multistep approach. Here, the radiances are sorted in ascending order. Then, unique radiance values are extracted and ascendingly sorted. Next, all adjacent differences are extracted, i.e. the second unique value is subtracted from the first one, the third unique value from the second one and so on. Then, the probability distribution of these differences is estimated by a histogram. The first frequency category (first bin) contains the smallest difference of unique values. The smallest difference is given as the minimum of all differences of this bin and represents the slope times the smallest difference of unique values (SDUV) of a perfectly calibrated band. The SDUV can be considered equivalent t0 the spectral detector resolution of the considered band. To estimate the slope, it is now necessary to assess the SDUV. This can be straightforwardly performed by computing the median of all binned differences. After dividing this smallest difference by the SDUV the slope for this band and detector is recovered. This is performed for each band and detector. After obtaining the slope coefficients the applicability is validated. This is performed by considering adjacent detector columns. For this, the shapes of the histograms of adjacent columns are inspected. If the number of frequency categories and the positions of the maxima are not equal, then the slope reduction is applied for the considered column. This evaluation bases on the assumption that significant different slopes of similar and adjacent detectors cause stretches (broadening) and shifts in the histogram since considered columns mostly cover the same regions and the related point spread functions (PSF) of each detector are stable during image acquisition and, hence, contribute to their neighbouring pixels the same fraction of their center pixel. In presence of c_0 offset miscalibration these offsets are reduced concurrently to c_0/c_1. Subsequently, SNR is computed to indicate whether previous operation is necessary or not. However, radiometric rescaling is then applied to reduce uncertainties in the estimation of SDUV (see section 3.5).

3.4 Linear c_0 reduction

In the following it is assumed that the thermally induced offset is constant during one image acquisition and that homogeneous regions are spectrally homogeneous. It follows from this that the offset of one detector element and wavelength contributes the same fraction to all pixels of one detector column and wavelength. Hence, spectral homogeneous regions that appear spectrally different indicate c_0 miscalibration if linear c_1 or nonlinear $c_{2..M}$ reductions were performed beforehand. To reduce c_0 miscalibration, it is necessary to spectrally compare adjacent image columns and to relate succeeding reduction to a predefined column (ROME uses per default the first column). In ROME the differences between adjacent columns are computed and binned in a histogram. Then, it is assumed that the bin (frequency category) with the highest frequency most likely contain the offset difference. To finally assess the offset difference, it is only necessary to average the differences of each bin by the median, to weight the bin according its frequency and to sum all weighted and averaged differences. After c_0 reduction a radiometric rescaling should be applied as in ROME to avoid erroneous radiometric levelling due to the used reference column. However, after applying an offset reduction, it is necessary to check whether this operation was necessary or not. In ROME this is performed by considering the evolution of the SNR.

3.5 Radiometric rescaling

Previous described approaches to correct data for miscalibration can change the mean radiation of a band that is only acceptable if the new mean is closer to a perfect calibrated band compared to the mean of the uncorrected band. This is not known yet and, hence, it is necessary to recover the physical meaning of such. A simple rescaling to the old maximum and minimum cannot be applied since it can be assumed that the old maximum and minimum are biased or erroneous due to miscalibration. In order to preserve the spectral characteristics a specific approach was proposed within the ROME framework as detection of lowest reduction zones. In this approach the correction vectors are inspected in a moving window. In each window the mean of the first and last reduction is rationed by the middle window reduction. After computing all windowed ratios the ratio that is closest to one is selected as reference. Then, the middle column of the reference is considered with regard to its maximum and minimum. The old maximum and minimum, i.e. before any reduction, is compared with the extrema of the reference. These are used to obtain linear transformation coefficients for the whole band that are subsequently applied.

3.6 Extended detrending

In Rogass, et al. (2011) a detrending approach is proposed that aims on the reduction of across track brightness gradients that are caused by offset reduction related frequency undershoots or by material, illumination and viewing geometry dependent surface responses on incident light. These undershoots have a medium frequency on average in comparison to the spatial distribution of the image content.

In ROME the detrending is realised per band by computing the median average of each column, by smoothing and mean normalising this column to its related average vector and by applying this vector on the image by row wise division.

However, lower frequencies are not considered in ROME as they can be perceived as broad brightness gradients. In this work, the new detrending approach is extended to capture lower frequency undershoots. For this, the column median per band of the uncorrected image and the corrected image is computed. This then gives one vector per band and image of the same length as the number of detectors. Each vector is then fitted to a second order polynomial with regard to least squares principles. Consequently, polynomial coefficients for each vector and image are obtained. The polynomial coefficients of the uncorrected image are subtracted from the coefficients of the corrected image. This gives differential coefficients for each band of the corrected image. After this an index vector is created that contains the same number of elements as detectors and consists of detector numbers (i.e. 0, 1, 2, 3... etc.). This could be considered as a x-vector. The x-vector is used to obtain functional values of the differential polynomials. This then gives the differential low frequency trend of this band with respect to the corrected and the uncorrected image. This trend is applied contrary to the detrending of ROME by row wise addition. Both the original detrending of ROME and this extension of the detrending enable a correction for medium and low frequency undershoots. A comparison of this approach and the originally proposed approach of ROME will be given in the results chapter.

3.7 Image quality metrics

In Rogass et al. (2011) several image quality metrics were combined to evaluate destriping results on the one hand and to avoid potential drawbacks associated with relying on a single type of evaluation on the other hand. In this work the same metrics are used. Those were the global Peak-Signal-to-Noise-Ratio (PSNR) (Rogass et al., 2010; Wang and Bovik, 2009), the global Shannon Entropy (Rogass et al., 2010, Frank and Smith, 2010) and the local Modified Structural Similarity Index (MSSIM) (Tsai and Chen, 2008; Wang and Bovik, 2009, Wang et al., 2004). In case of available ground truth as for the HyMAP scene the metrics were applied on the result and on ground truth. In case of missing ground truth the metrics were applied on both the input and the output, but can only be relatively considered.

4. Results and discussion

The ROME framework is the most recent approach to recover radiometric calibration in presence of miscalibration. In this work more tests were included to show that ROME is able to reduce miscalibration of broadly used sensors. The summarised results of Tab. 1 show how miscalibration was reduced that is detailed discussed per newly considered sensor in the next sections. All newly tested sensors were miscalibrated due to varying dark current.

Sensor	Scene	PSNR	Entropy	MSSIM	Average
APEX	1	4 %	19 %	1 %	8 %
ASTER	1	4 %	19 %	1 %	8 %
CHRIS	1	1 %	0 %	3 %	1 %
HyMAP	SNR=7.6	-5 %	4%	6 %	5 %
	SNR=76	0 %	3 %	5 %	3 %
	SNR=760	0 %	2 %	5 %	2 %
AISA[1,2]	1	-2 %	9 %	8 %	5 %
EnMAP[1,2]	1	2 %	8 %	7%	6 %
Grey images[1,2]	3,4	4 %	4 %	2 %	3 %
Hyperion[1,2]	1	2 %	5 %	7 %	5 %

[1] compared to ground truth; [2] from Rogass et al. (2011)

Table 1. Destriping results

4.1 Grey valued images

The grey valued images that have been selected for testing in Rogass et al. (2011) cover a broad range of spectral and spatial image properties. In this work 2 out of 4 of the test images were selected due to their similar spatial and spectral distributions compared to remote sensing scenes. The 'Aerial' image is characterised by leptokurtic grey value distribution. The 'Sailboat on lake' image has a balanced grey value distribution and edge quantity. With regard to Rogass et al. (2011) ROME achieved a destriping accuracy of 97 % (compare Tab. 1 and Fig. 3) for the two grey test images. As perceptible in Fig. 3 all stripes were removed and the results differ from ground truth (Fig. 1c and d) only by 3% on average (Tab. 1).

4.2 Artificial striped HyMAP

The HyMAP whiskbroom sensor was three times differently offset striped, ROME destriped and the results were evaluated based on the metrics of section 3.7. The offset stripes were generated as described in Rogass et al. (2011) and scaled to achieve an overall SNR of 7.6, 76 and 760. The offset stripe type was selected since this type is most common to broadly used pushbroom sensors. However, about 97 % of a perfect calibration could be recovered (compare Tab. 1). Hence, the accuracy assumption of Rogass et al. (2011) that 97 % of a perfect calibration can be recovered by ROME is confirmed. With regard to the results visually presented in Fig. 4 the stripes were completely removed.

Fig. 3. Striped (left) and destriped images (right) for a)'Aerial' and b) 'Sailboat on lake'

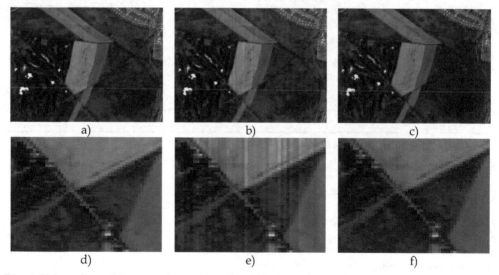

Fig. 4. False coloured image subset of band 30 (874 nm) of a HyMAP scene (subset a and zoom d), striped representation with a SNR of 7.6 (subset b and zoom e) and the ROME result adaptively detrended (subset c and zoom f)

Uncertainties remain in the assessment of the true radiometric scale as well as in the correct trend. This is visualised in Fig. 5. Considering both the transect and the spectral profile of Fig. 5 leads to the perception that small differences between ground truth and the destriping result persist. These differences approximately amounts 3% due to Tab. 1. This underlines the robustness of the ROME approach and contemporary shows that miscalibration can be efficiently suppressed.

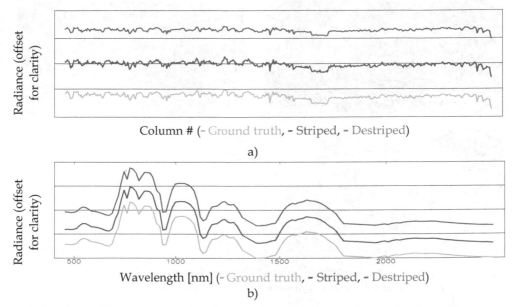

Fig. 5. Random arbitrary transect a) and spectral profile for a random point due to the subsets of Fig. 4 a), b) and c)

4.3 ASTER

The ASTER sensor was selected for destriping since it has broader bands as an typical hyperspectral sensor and the potential miscalibration is often underestimated in the literature. However, the visible and near infrared bands were selected since these bands were mostly preceptible miscalibrated as exemplarily shown in Fig. 6. With regard to the results of Tab. 1 the destriping of the ASTER scene improved the radiometric calibration by 8 % on average. That is significant in comparison to the CHRIS/Proba related destriping results. As perceptible in Fig. 6 all stripes were removed.

As shown in Fig. 6 and 7 miscalibration is mostly visually perceptible in contrary to arbitrary transects as presented in Fig. 7a). However, the ROME framework and the adaptive detrending reduced the miscalibration. In consequence, the spectral profile has changed as given in Fig. 7 b). Contrary to airborne sensors miscalibrations of satellite sensors such as ASTER slowly vary over time. It follows from this that correction sets obtained by the ROME framework can be reused for scenes that are timely close.

4.4 CHRIS/Proba

As shown in Fig. 8 the test scene acquired by the CHRIS sensor is well calibrated. However, remaining miscalibration is visually perceptible as given in Fig. 8 c).

With regard to Tab. 1 ROME improved the radiometric calibration by 1 % on average. This shows on the one hand that the scene of this sensor was well calibrated and on the other hand that ROME is also able to detect and to reduce small variations of miscalibrations.

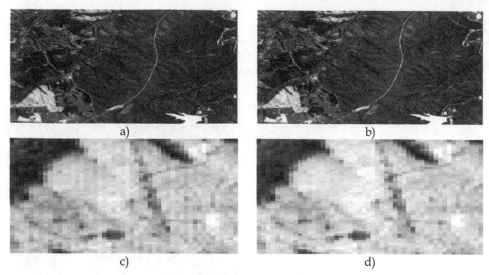

Fig. 6. False coloured image subset of band 3 (807 nm) of a striped ASTER scene (subset a and zoom c) and the ROME result adaptively detrended (subset b and zoom d)

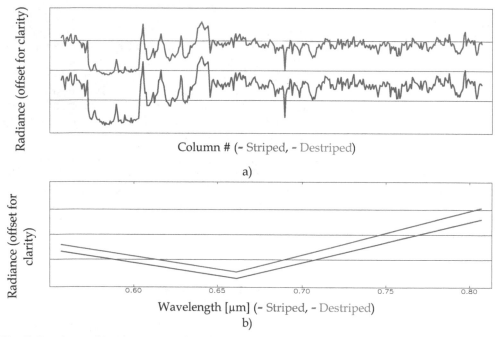

Fig. 7. Random arbitrary transect a) and spectral profile for a random point due to the subsets of Fig. 6 a) and b)

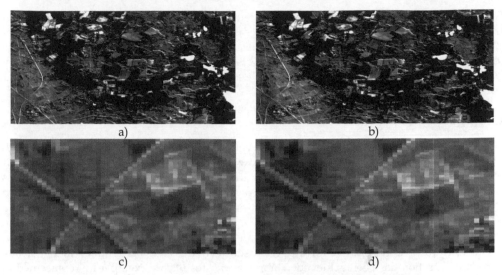

Fig. 8. False coloured image subset of band 44 (803.8 nm) of a striped CHRIS/Proba scene (subset a and zoom c) and the ROME result adaptively detrended (subset b and zoom d)

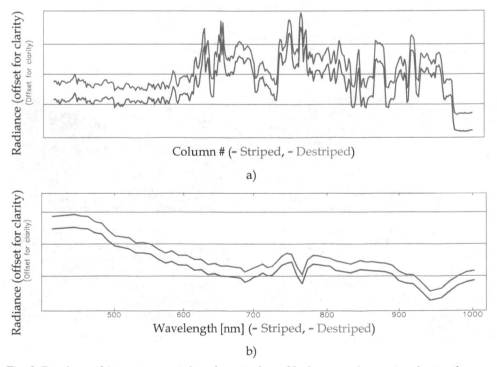

Fig. 9. Random arbitrary transect a) and spectral profile for a random point due to the subsets of Fig. 8 a) and b)

The transect as well as the spectral profile given in Fig. 9 show that ROME preserved spatial and spectral shapes. Contrary to ASTER it appears that the ROME destriping of CHRIS/Proba scenes is only necessary if succeeding processing consider adjacent image columns. In relation to Rogass et al. (2011) 97 % of a perfect calibration can be recovered by ROME. It follows from this that the decision whether ROME is applied on CHRIS/Proba or not should be application driven.

4.5 APEX

The APEX sensor belongs to the recently developed pushbroom sensors and offers a high SNR for a broad set of applications. However, as most pushbroom sensors APEX acquisitions also show perceptible variations in dark current as offset stripes although it is well calibrated like CHRIS/Proba. These stripes are difficult to be detected due to the high SNR of APEX and to the overall low contribution of miscalibration to image spectra. To additionally test the new detrending approach, a subset of a scene (400 lines) was used. In consequence, the results of Tab. 1 that show an overall improvement of calibration of about 8 % are not fully representative for the APEX sensor. In this case it is assumed that 97 % of a perfect calibration has been achieved. The respective results are exemplarily represented in Fig. 10 and 11. Comparing the along track transect of Fig. 11 a) and the spectral profile of Fig. 11 b) with the false coloured image representations of Fig. 10 it appears that changes of spectra are mostly visually perceptible. That supports the assumption that APEX acquisitions are not dominated by dark current variations contrary to Hyperion or AISA DUAL. The assumption that potential frequency undershoots caused by, e.g. offset reductions, are minimised by the new detrending approach is also supported (compare also next chapter).

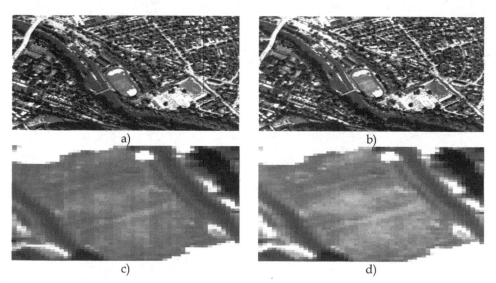

Fig. 10. False coloured image subset of band 19 (557.3 nm) of a striped APEX scene (subset a and zoom c) and the ROME result adaptively detrended (subset b and zoom d)

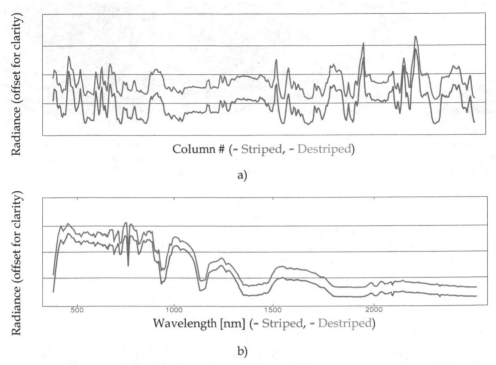

Column # (- Striped, - Destriped)

a)

Wavelength [nm] (- Striped, - Destriped)

b)

Fig. 11. Random arbitrary transect a) and spectral profile for a random point due to the subsets of Fig. 10 a) and b)

4.6 Results for extended detrending

The ROME framework as proposed in Rogass et al. (2011) has limited facilities for short scenes. In this work the impact of short scenes is inspected and an extension to its detrending proposed. Since the effect varies from scene to scene and sensor to sensor it is not possible to quantify the impact. To qualify the impact of short scenes on ROME, one artifically offset striped HyMAP scene subset (SNR=7.6) was destriped. Then, the result was ROME detrended and detrended by the nex approach. The respective results are given Fig. 12 and 13. As perceptible in Fig. 12 b) and e) compared to Fig. 12 c and f significant reduction related brightness gradients are significantly reduced by the new approach.

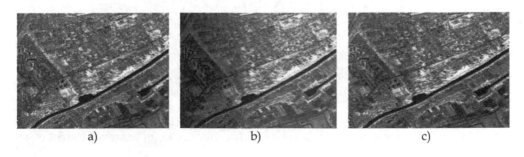

a)   b)   c)

d)   e)   f)

Fig. 12. False coloured, small image subset of band 30 (874 nm) of a HyMAP scene (subset a and zoom d) that war artificially offset striped (SNR=7.6), ROME result (subset b and zoom e) and the ROME result adaptively detrended (subset c and zoom f)

The across track transect as well as the spectral profile given in Fig. 13 clearly show the impact of the detrending on the spectral scale. Comparing the old detrending approach with the new detrending approach leads to the perception that the new detrending preserves the spectral profile in both directions the spatial domain - across track (correction direction) and the spectral domain – along the spectrum.

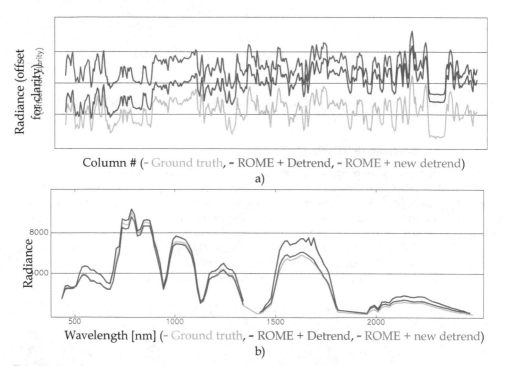

Fig. 13. Random arbitrary transect a) and spectral profile for a random point due to the subsets of Fig. 10 a) and b)

It follows from this that relatively short scenes are more difficult to correct as long scenes. In Rogass et al. (2011) it was assumed that the ROME correction facilities are dependent on the along track dimension. This is supported and can be clearly demonstrated, e.g. by transects and spectral profiles of corrected short scenes as presented in Fig. 12 and 13. The subsets for detrending comparisons had a size of 400 lines.

5. Conclusions

Pushbroom sensors must be carefully calibrated and miscalibrations aggravate succeeding operations such as atmospheric correction (Richter, 1997), classification and segmentation (Datt et al., 2003). Therefore, it is necessary to efficiently reduce them. The ROME framework and the extended detrending proposed in this work significantly reduce miscalibrations of any type. Like other methods there are also limitations. These limitations mostly relate to offset and nonlinear reductions, not the linear slope reduction.

However, a calibration recovery rate of about 97 % still remains uncertainties. High spatial densities of translucent objects such as trees reduce offset reduction facilities and should be excluded beforehand. Tests with different data sets also showed that dense haze or clouds may hinder offset reduction. These effects can be minimised by destriping subsets and by applying estimated correction coefficients on the whole image. In case of clouds or dense haze a reference column for offset reduction that is haze or cloud free is suggested.

With regard to tests of Rogass et al. (2011) and tests performed for this work it can be assumed that the ROME framework is capable to reduce miscalibrations for most pushbroom sensors. With regard to the high processing speed and the freedom of parameters it can be operationally used. The nonlinear correction has to be improved but represents the current state of the art method as the other methods implemented in ROME. However, further research is necessary. This is particularly applicable for high frequency undershoots that are currently not considered.

6. References

Atkinson, P.M.; Sargent, I.M.; Foody, G.M.; Williams, J. Interpreting Image-Based Methods for Estimating the Signal-to-Noise Ratio. Int. J. Rem. Sens. 2005, 26, 5099–5115.

Barducci, A.; Castagnoli, F.; Guzzi, D.; Marcoionni, P.; Pippi, I.; Poggesi, M. Solar Spectral Irradiometer for Validation of Remotely Sensed Hyperspectral Data. Appl. Opt. 2004, 43, 183–195.

Barnsley, M. J., Allison, D., Lewis, P. 1997. On the information content of multiple view angle (MVA) images. International Journal of Remote Sensing, 18:1936- 1960.

Biggar, S.; Thome, K.; Wisniewski, W. Vicarious Radiometric Calibration of EO-1 Sensors by Reference to High-Reflectance Ground Targets. IEEE Trans. Geosci. Rem. Sens.2003, 41, 1174–1179.

Bindschadler, R.; Choi, H. Characterizing and Correcting Hyperion Detectors Using Ice-Sheet Images. IEEE Trans. Geosci. Rem. Sens. 2003, 41, 1189–1193.

Bouali, M.; Ladjal, S. A Variational Approach for the Destriping of Modis Data. In IGARSS 2010: Proceedings of the IEEE International Geoscience and Remote Sensing Symposium, Honolulu, Hawaii, 25–30 July, 2010; pp. 2194–2197.

Box, G.; Muller, M. A Note on the Generation of Random Normal Deviates. Ann. Math. Stat. 1958, 29, 610–611.

Bruegge, C.; Diner, D.; Kahn, R.; Chrien, N.; Helmlinger, M.; Gaitley, B.; Abdou, W. The Misr Radiometric Calibration Process. Rem. Sens. Environ. 2007, 107, 2–11.

Brunn, A.; Fischer, C.; Dittmann, C.; Richter, R. Quality Assessment, Atmospheric and Geometric Correction of Airborne Hyperspectral Hymap Data. In Proceedings of the 3rd EARSeL Workshop on Imaging Spectroscopy, Herrsching, Germany, 13–16 May 2003; pp. 72–81.

Canny, J. A Computational Approach to Edge Detection. IEEE Trans. Pattern Anal. Mach. Intell. 1986, 8, 679–698.

Carfantan, H.; Idier, J. Statistical linear destriping of satellite-based pushbroom-type images. IEEE Trans. Geosci. Rem. Sens. 2010, 48, 1860–1871.

Cavalli, R.; Fusilli, L.; Pascucci, S.; Pignatti, S.; Santini, F. Hyperspectral Sensor Data Capability for Retrieving Complex Urban Land Cover in Comparison with Multispectral Data: Venice City Case Study (Italy). Sensors 2008, 8, 3299–3320.

Chander, G.; Markham, B.; Helder, D. Summary of Current Radiometric Calibration Coefficients for Landsat MSS, TM, ETM+, and EO-1 ALI Sensors. Rem. Sens. Environ. 2009, 113, 893–903.

Cocks, T., Jenssen, R., Stewart, A., Wilson, I., and Shields, T., 1998, The HyMap airborne hyperspectral sensor: the system, calibration and performance, First EARSeL Workshop on Imaging Spectroscopy, 6-8 Oct. 1998, Zurich, Switzerland, pp. 37-42.

Datt, B.; McVicar, T.R.; van Niel, T.G.; Jupp, D.L.B.; Pearlman, J.S. Preprocessing EO-1 Hyperion Hyperspectral Data to Support the Application of Agricultural Indexes. IEEE Trans. Geosci. Rem. Sens. 2003, 41, 1246–1259.

Dell'Endice, F. Improving the Performance of Hyperspectral Pushbroom Imaging Spectrometers for Specific Science Applications. In ISPRS 2008: Proceedings of the XXI Congress: Silk Road for Information from Imagery: The International Society for Photogrammetry and Remote Sensing, 3–11 July, Beijing, China, 2008; pp. 215–220.

Dell'Endice, F.; Nieke, J.; Koetz, B.; Schaepman, M.E.; Itten, K. Improving Radiometry of Imaging Spectrometers by Using Programmable Spectral Regions of Interest. ISPRS J. Photogramm. Rem. Sens. 2009, 64, 632–639.

Frank, S.; Smith, E. Measurement Invariance, Entropy, and Probability. Entropy 2010, 12, 289–303.

Gao, B.-C. An Operational Method for Estimating Signal to Noise Ratios from Data Acquired with Imaging Spectrometers. Rem. Sens. Environ. 1993, 43, 23–33.

García, J.; Moreno, J. Removal of Noises in CHRIS/Proba Images: Application to the SPARC Campaign Data. In Proceedings of the 2nd CHRIS/Proba Workshop, ESA/ERSIN, Frascati, Italy, 28–30 April, 2004; pp. 29–33.

Gómez-Chova, L.; Alonso, L.; Guanter, L.; Camps-Valls, G.; Calpe, J.; Moreno, J. Correction of Systematic Spatial Noise in Push-Broom Hyperspectral Sensors: Application to CHRIS/Proba Images. Appl. Opt. 2008, 47, F46-F60.

Guanter, L.; Segl, K.; Kaufmann, H. Simulation of optical remote-sensing scenes with application to the enmap hyperspectral mission. IEEE Trans. Geosci. Rem. Sens. 2009, 47, 2340–2351.

Haralick, R.M.; Sternberg, S.R.; Zhuang, X. Image Analysis Using Mathematical Morphology. IEEE Trans. Pattern Anal. Mach. Intell. 1987, 9, 532–550.

Itten, K. I.; Dell'Endice, F.; Hueni, A.; Kneubühler, M.; Schläpfer, D. ; Odermatt, D.; Seidel, F.; Huber, S.; Schopfer, J.; Kellenberger, T.; Bühler, Y.; D'Odorico, P.; Nieke, J.; Alberti, E.; Meuleman, K. APEX - the Hyperspectral ESA Airborne Prism Experiment. Sensors 2008, 8, 6235-6259.

Kaufmann, H.; Segl, K.; Guanter, L.; Förster, K.P.; Stuffler, T.; Müller, A.; Richter, R.; Bach, H.; Hostert, P.; Chlebek, C. Environmental Mapping and Analysis Program (EnMAP) — Recent Advances and Status. In IGARSS 2008: Proceedings of the IEEE International Geoscience and remote Sensing Syposium, 7–11 July, Boston, MA, USA, 2008; pp. IV-109-IV-112.

Le Maire, G.; François, C.; Soudani, K.; Berveiller, D.; Pontailler, J.-Y.; Bréda, N.; Genet, H.; Davi, H.; Dufrêne, E. Calibration and Validation of Hyperspectral Indices for the Estimation of Broadleaved Forest Leaf Chlorophyll Content, Leaf Mass Per Area, Leaf Area Index and Leaf Canopy Biomass. Rem. Sens. Environ. 2008, 112, 3846–3864.

Liu, B.; Zhang, L.; Zhang, X.; Zhang, B.; Tong, Q. Simulation of EO-1 Hyperion Data from ALI Multispectral Data Based on the Spectral Reconstruction Approach. Sensors 2009, 9, 3090–3108.

Oliveira, P.; Gomes, L. Interpolation of Signals with Missing Data Using Principal Component Analysis. Multidimens. Syst. Signal Process. 2010, 21, 25–43.

Oppelt, N.; Mauser, W. The Airborne Visible/Infrared Imaging Spectrometer Avis: Design, Characterization and Calibration. Sensors 2007, 7, 1934–1953.

Richter, R. Correction of Atmospheric and Topographic Effects for High Spatial Resolution Satellite Imagery. Int. J. Rem. Sens. 1997, 18, 1099–1111.

Rogass, C.; Itzerott, S.; Schneider, B.; Kaufmann, H.; Hüttl, R. Edge Segmentation by Alternating Vector Field Convolution Snakes. Int. J. Comput. Sci. Netw. Secur. 2009, 9, 123–131.

Rogass, C.; Itzerott, S.; Schneider, B.; Kaufmann, H.; Hüttl, R. Hyperspectral Boundary Detection Based on the Busyness Multiple Correlation Edge Detector and Alternating Vector Field Convolution Snakes. ISPRS J. Photogramm. Rem. Sens. 2010, 55, 468–478.

Rogass, C.; Spengler, D.; Bochow, M.; Segl, K.; Lausch, A.; Doktor, D.; Roessner, S.; Behling, R.; Wetzel, H.-U.; Kaufmann, H. Reduction of Radiometric Miscalibration — Applications to Pushbroom Sensors. Sensors 2011, 11, 6370-6395.

Segl, K.; Guanter, L.; Kaufmann, H.; Schubert, J.; Kaiser, S.; Sang, B.; Hofer, S. Simulation of Spatial Sensor Characteristics in the Context of the EnMAP Hyperspectral Mission. IEEE Trans. Geosci. Rem. Sens. 2010, 48, 3046–3054.

Shen, H.F.; Ai, T.H.; Li, P.X. Destriping and Inpainting of Remote Sensing Images Using Maximum a-Posteriori Method. In ISPRS 2008: Proceedings of the XXI Congress: Silk Road for Information from Imagery: The International Society for Photogrammetry and Remote Sensing, 3–11 July, Beijing, China, 2008; pp. 63–70.

Simpson, J.J.; Gobat, J.I.; Frouin, R. Improved Destriping of Goes Images Using Finite Impulse Response Filters. Rem. Sens. Environ. 1995, 52, 15–35.

Simpson, J.J.; Stitt, J.R.; Leath, D.M. Improved Finite Impulse Response Filters for Enhanced Destriping of Geostationary Satellite Data. Rem. Sens. Environ. 1998, 66, 235–249.

Spectral Imaging Ltd. Aisa Dual, 2nd Version. Available online: http://www.specim.fi/media/aisa-datasheets/dual_datasheet_ver2-10.pdf (accessed on 5 January 2011).

Tsai, F.; Chen, W. Striping Noise Detection and Correction of Remote Sensing Images. IEEE Trans. Geosci. Rem. Sens. 2008, 46, 4122–4131.

Ungar, S.G.; Pearlman, J.S.; Mendenhall, J.A.; Reuter, D. Overview of the Earth Observing One (EO-1) Mission. IEEE Trans. Geosci. Rem. Sens. 2003, 41, 1149–1159.

Wang, Z.; Bovik, A.C. Mean Squared Error: Love It or Leave It? A New Look at Signal Fidelity Measures. IEEE Signal Process. Mag. 2009, 26, 98–117.

Wang, Z.; Bovik, A.C.; Sheikh, H.R.; Simoncelli, E.P. Image Quality Assessment: From Error Visibility to Structural Similarity. IEEE Trans. Image Process. 2004, 13, 600–612.

Weber, A. The USC-SIPI Image Database; Technical Report, University of Southern California, Signal and Image Processing Institute: Los Angeles, CA, USA, 1997.

Xiong, X.; Barnes, W. An Overview of Modis Radiometric Calibration and Characterization. Adv. Atmos. Sci. 2006, 23, 69–79.

Yamaguchi, Y.; Kahle, A.B.; Tsu, H.; Kawakami, T.; Pniel, M. Overview of Advanced Spaceborne Thermal Emission and Reflection Radiometer (ASTER). IEEE Trans. Geosci. Remote Sensing 1998, 36(4), 1062-1071.

8

Energy Efficient Data Acquistion in Wireless Sensor Network

Ken C. K. Lee[1], Mao Ye[2] and Wang-Chien Lee[2]
[1]*Department of Computer and Information Science,*
University of Massachusetts Dartmouth, North Dartmouth,
[2]*Department of Computer Science and Engineering,*
The Pennsylvania State University, University Park,
USA

1. Introduction

Wireless sensor network (or sensor network, for brevity in the following) comes into practice, thanks to the recent technological advancement of embedded systems, sensing devices and wireless communication. A typical sensor network is composed of a number of wirelessly connected sensor nodes distributed in a sensed area. In the network, sensor nodes sense their surroundings and record sensed readings. The sensed readings of individual sensor nodes are then collected to present the measurement of an entire sensed area. In many fields including but not limit to, military, science, remote sensing Vasilescu et al. (2005), industry, commerce, transportation Li et al. (2011), public security Faulkner et al. (2011), healthcare and so on, sensor networks are recognized as important sensing, monitoring and actuation instruments. In addition, many off-the-shelf sensor node products Zurich (n.d.) and supporting software such as TinyOS Group (n.d.) are available in the market. Now sensor network application development is much facilitated. Many sensor networks are anticipated to be deployed soon.

Over those years, the computational capability and storage capacity of sensor nodes have been considerably improving. Yet, the improvement of battery energy is relatively small. Since battery replacement for numerous deployed sensor nodes is extremely costly and even impossible in hostile environments, battery energy conservation is a critical issue to sensor networks and their applications. Accordingly, how to effectively save battery energy is a challenge to researchers from academia, government agencies and industries. One common practice is to keep sensor nodes in sleep mode whenever they are not in use. During sleep mode, some hardware components of sensor nodes are turned off to minimize energy consumption. For instance, MICAz needs only $1\mu A$ when wireless interface is off and less than $15\mu A$ for processor in sleep mode Musaloiu-Elefteri et al. (2008). Besides, wireless communication is very energy consuming. For instance, MICAz consumes 17.4mA and 19.7mA in data sending and receiving, respectively, whereas it only needs 8mA for computation when its wireless interface and processor are on. Thus, reducing the amount of data transmitted between sensor nodes is another important means to save battery energy.

In many sensor network applications, data acquisition that collects sensed readings from remote sensor nodes is an essential activity. A primitive approach for data acquisition can be collecting all raw sensed readings and maintaining them in a data repository for centralized processing. Alternatively, a large volume of raw sensed readings are streamed to a processing site where analysis and data processing are directly applied on streamed sensor readings Madden & Franklin (2002). However, costly wireless communication can quickly use up sensor nodes' battery energy. In other words, such a centralized approach is not energy efficient and thus undesirable in practice. As in the literature, a lot of original ideas and important research results have been developed for energy efficient data acquisition. Among those, many new techniques have been developed based on the idea of in-network query processing. Through in-network query processing, queries are delivered into sensor networks and sensor nodes evaluate the queries locally. By doing so, (partial) query results are transmitted instead of raw sensed readings. Since (partial) query results are in smaller size than raw sensed readings, energy cost can be effectively saved. Subject to the types of queries and potential optimization opportunities, various in-network query processing techniques have been developed and reported in the literature.

In this chapter, we review the main concepts and ideas of many representative research results on in-network query processing, which include some of our recent works such as itinerary-based data aggregation Xu et al. (2006), materialized in-network view Lee et al. (2007), contour mapping engine Xu et al. (2008) and in-network probabilistic minimum value search Ye, Lee, Lee, Liu & Chen (to appear). As briefly described, itinerary-based data aggregation is a new access method that navigates query messages among sensor nodes to collect/aggregate their sensed readings. Materialized in-network view is a novel data caching scheme that maintains (partial) query results in queried sensor nodes. Then, subsequent queries issued by different base stations can access cached results instead of traversing query regions from scratch to determine query results. Contour mapping engine derives fairly accurate contour line segments using data mining techniques. Besides, only the coefficients of equations representing contour line segments, which are very compact, are transmit. Finally, probabilitistic minimum value search is one of recent efforts in probabilistic sensed data aggregation. It finds the possible smallest sensed reading values in a sensor network.

The details of those works will be discussed in the following sections. First of all, we present a system model that our reviewed research results are based upon. Then, we discuss research results in in-network data aggregation and in-network data caching as well as in-network contour map computation. We further discuss recent results on in-network probabilistic data aggregation. Last but not least, we summarize this chapter and discuss some future research directions.

2. System model

Without loss of generality, a sensor network is composed of a number of battery powered stationary sensor nodes deployed over a sensed area. The spatial deployment of sensor nodes in a target sensed area is one of the research problems in sensor networks; and many research works (e.g. Bojkovic & Bakmaz (2008)) were proposed to maximize the area coverage by a given quantity of sensor nodes while providing required network connectivity among sensor nodes. The issue of sensor node deployment is usually considered to be independent from others. As will be discussed in the following, research works on data acquisition mostly assume that sensor networks are already set up and all sensor nodes are with identical hardware configurations.

In a typical sensor network, some senor nodes in the sensor network are directly connected to computer terminals; and they are called *base stations*. Through base stations, computer terminals can issue commands to administer sensor nodes and collect their sensed readings. Besides, all sensor nodes are wirelessly connected, e.g., MICAz uses 2.4GHz IEEE 802.15.4 radio. That means messages are all sent through wireless broadcast. When a node delivers a message, other sensor nodes within its radio coverage range can receive the message. Messages can be conveyed transitively from a sender sensor node to a distant target receiver node Xu et al. (2007). On the other hand, because of shared radio frequencies, simultaneous messages from closely located sensor nodes may lead to signal interference. Moreover, due to ad hoc connectivity and sensor node failure, which is common in practice, connections among sensor nodes are mostly transient and unreliable. Thus, other than regular data messages, every sensor node periodically broadcasts a special message called *beacon* to indicate its liveness to its neighboring sensor nodes. Also, data messages are sent through multiple paths from a sender sensor node towards a destination to deal with possible message loss Xu et al. (2007). As a result, those extra messages incur additional energy costs.

To save battery energy, sensor nodes stay in sleep mode for most of the time; and each of them periodically wakes up to sense its surrounding and record its measurements as sensed readings. For data acquisition, an entire sensor network (i.e., a set of sensor nodes N) presents a set of sensed reading values V, notationally, $V = \{v_n \mid n \in N\}$ where v_n is a sensed reading value provided by a sensor node n. Based on V, data analysis is conducted to understand the entire sensed area. As already discussed, it is very costly to collect V from all sensor nodes. Accordingly, some research results were reported in the literature exploring techniques to collect a subset of sensed readings $V' (\subset V)$ from a subset of sensor nodes N' $(\subset N)$, while collected readings may only provide approximate analytical results. The following are two sorts of techniques. Sampling is the first technique that sensed readings are only collected from some (randomly) selected sensor nodes Biswas et al. (2004); Doherty & Pister (2004); Huang et al. (2011). Those unselected sensor nodes do not need to provide their sensed readings. The sampling rate is adjustable according to the energy budget. The second technique is based on a certain prediction model Silberstein et al. (2006) that, some sensed readings can be omitted from being sent as long as they can be (approximately) predicted according to other sensed readings, which can be from some neighboring sensor nodes, or from the previous sensed reading values of the same sensor nodes. Meanwhile, another important research direction for energy efficient data acquisition based on in-network query processing Hellerstein et al. (2003) has been extensively studied; and we shall review some of the representative works in the coming four sections.

3. In-network data aggregation

Data aggregation is often used to summarize a large dataset. With respect to all sensed readings V from all sensor nodes N, an aggregate function f is applied on V to obtain a single aggregated value, i.e., $f(V)$. Some commonly used aggregate functions include SUM, COUNT, MEAN, VARIANCE, MAX and MIN etc. Aggregated data can provide a very small summary of sensed readings (e.g., the highest, average and lowest temperature) in a sense area. In many situations, it can be sufficient for scientists to know about a remote sensed area. Besides, aggregated data is usually small to transmit and data aggregation is not very computationally expensive for sensor nodes to perform so that in-network data aggregation is very suitable to sensor networks. In the following, we discuss two major strategies, namely, *infrastructure-based approaches* and *itinerary-based approaches*, for in-network data aggregation.

3.1 Infrastructure-based data aggregation

As their name suggests, infrastructure-based approaches build certain routing structures among sensor nodes to perform in-network data aggregation. TAG Madden et al. (2002) and COUGAR Yao & Gehrke (2003) are two representative infrastructure-based approaches. They both form a routing tree to disseminate a query and to derive aggregated sensed readings in divide-and-conquer fashion. The rationale behind these approaches are two ideas. First, some aggregate functions f are decomposable so that $f(V)$ can be transformed to $f(f(V_1), f(V_2), \cdots f(V_x))$, where $V_1, V_2, \cdots V_x$ are sensed reading values from x disjointed subsets of sensor nodes and the union of all of them equals V, and f can be applied to readings from individual subsets of sensor nodes and to their aggregated readings. For example, $SUM(V)$, where SUM adds all sensed reading values, can be performed as $SUM(SUM(V_1), SUM(V_2), \cdots SUM(V_x))$. Second, the connections among sensor nodes can be organized as a tree topology, in which the root of any subtree that covers a disjointed subset of some sensor nodes can carry out local aggregation on data from its descendant nodes. In other words, in-network data aggregation incrementally computes aggregated values at different levels in a routing tree.

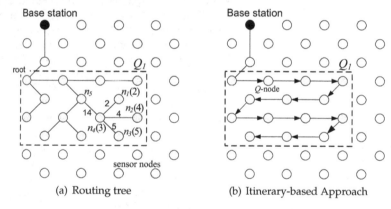

(a) Routing tree   (b) Itinerary-based Approach

Fig. 1. Strategies for in-network data aggregation

Figure 1(a) exemplifies a routing tree formed for data aggregation. In brief, upon receiving a SUM query for the total of sensed reading values from its connected computer terminal, a base station disseminates the query to sensor nodes within a specified queried region. The specified queried region can be a small area or an entire sensed area. With the queried region, sensor nodes join the routing tree when they receive the query. A node becomes the parent node of its neighboring nodes in a routing tree if those nodes receive the query from it. In a routing tree, the first queried node within the region serves as the root. Meanwhile, every non-root tree node should have another sensor node as its parent node, and non-leaf nodes are connected to some other nodes as their child nodes.

After the tree is built, data aggregation starts from leaf nodes. The leaf nodes send their sensed reading values to their parent nodes. Thereafter, every non-leaf node derives an aggregated value based on (aggregated) sensed reading values received from its child nodes and its own sensed reading value. As shown in Figure 1(a), some leaf nodes n_1, n_2, n_3 first send their reading values of 2, 4 and 5, respectively, to their parent node n_4. Then, n_4 calculates the sum of their values and its own sensed reading values of 3, i.e., 14, and propagates it to its parent node n_5. Eventually, the root derives the final sum among all sensor nodes in the region and reports it to the base station.

3.2 Itinerary-based data aggregation

The infrastructure-based approaches relies on an infrastructure to perform in-network data aggregation, incurring two rounds of messages for both query dissemination and data collection. However, in presence of sensor node failure, queries and aggregated sensed readings would be lost making these approaches not very robust and reliable. Some additional research works Manjhi et al. (2005) were proposed to improve the robustness and reliability of routing trees by replicating aggregated values and sending them through different paths towards the root. However, it incurs extra data communication cost. To save the quantity of messages, we have recently developed itinerary-based data aggregation Xu et al. (2006).

The basic idea of itinerary-based data aggregation is to navigate a query among sensor nodes in a queried region as illustrated in Figure 1(b). In every step, a query message that carries both a query specification and an immediate query result is strategically sent from one sensor node to another along a designed space filling path called *itinerary*. The width of an itinerary is bounded by a maximum radio transmission range. Sensor nodes participating in forwarding a query message are called Q-nodes. After it receives a query message, a Q-node asks its neighboring nodes for their sensed readings. Then, the Q-node incorporates all received sensed readings and its own reading into the immediate query result. Thereafter, it forwards the query message with a new intermediate query result to a succeeding Q-node. Here, the succeeding Q-node is chosen by the current Q-node. If a Q-node fails, its preceding Q-node can detect it and re-propagates the query message to another sensor node as a replacement Q-node. As such, the itinerary can be resumed from that new Q-node. The evaluation of a query completes when a specified region is completely traversed. Finally, a query result is returned to the base station.

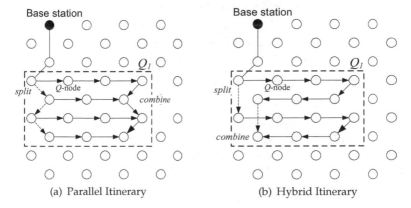

(a) Parallel Itinerary   (b) Hybrid Itinerary

Fig. 2. Parallel and hybrid itinerary

On the other hand, the length of an itinerary directly affects the query processing time. A single itinerary takes a very long processing time, especially in a large query region. Thus, as opposed to single itinerary as shown in Figure 1(b), parallel itinerary has been developed to improve query processing time. As depicted in Figure 2(a), an itinerary is split into four threads scanning four rows in a region. Their immediate query results are then aggregated at the end of the rows. However, wireless signal from two adjacent threads may lead to signal interference, message loss and finally data retransmission. As a result, longer time and more energy are consumed. To address this issue, a hybrid itinerary has been derived accordingly. Here, a query region is divided into several sections that contain multiple rows. Inside each section, a single itinerary scans all the rows. For instance, as in Figure 2(b), a query region is partitioned into two sections, each covering two rows. Within each section, a sequential itinerary is formed. Now, because of wider separation, the impact of signal interference is minimized while a higher degree of parallelism is achieved, compared with single itinerary.

Through simulation, our developed itinerary-based approach is demonstrated outperforming infrastructure-based approaches Xu et al. (2006). Besides, the idea of itinerary-based in-network query processing has also been adopted for other types of queries and applications such as tracking nearest neighbor objects Wu et al. (2007).

4. In-network data caching

Data caching is widely used in distributed computer systems to shorten remote data access latency. In sensor networks, data caching has one more important benefit that is saving communication energy cost. Many existing research works focused on strategies of replicating frequently accessed sensed readings in some sensor nodes closer to base stations Ganesan et al. (2003); Liu et al. (2004); Ratnasamy et al. (2002); Sadagopan et al. (2003); Shakkottai (2004); Zhang et al. (2007). In presence of multiple base stations, a research problem of finding sensor nodes for caching sensed readings is formulated as determining a Steiner tree in a sensor network Prabh & Abdelzaher (2005). In a graph, a Steiner tree is a subgraph connecting all specified vertices and providing the smallest sum of edge distances Invanov & Tuzhilin (1994). By caching data in some sensor nodes as internal vertices (that connect more than one edge) in a Steiner tree, the communication costs between those sensor nodes providing sensed readings and base stations are guaranteed to be minimized.

On the other hand, existing data caching schemes do not support data aggregation. Accordingly, we have devised a new data caching scheme called *materialized in-network view* (MINV) to support SUM, AVERAGE, COUNT, VARIANCE aggregate functions Lee et al. (2007). Specifically, MINV maintains partially computed aggregated readings in some queried sensor nodes. Then, subsequent queries, which are issued by different base stations and which cover queried sensor nodes, can be fully or partially answered by cached results.

Figure 3(a) shows a motivating example of MINV. In the figure, a SUM query Q_1 adds up the sensed readings of all sensor nodes in a query region at time t_1. At a later times t_2 and t_3, two other SUM queries, Q_2 and Q_3, respectively, are issued to summarize readings from sensor nodes in two other queried regions overlapping Q_1's. Without cache, all queries are processed independently. Ideally, if Q_1's answer can be maintained and made accessible, Q_2 and Q_3 can be answered by some cached data to save the energy costs of an entire sensor network.

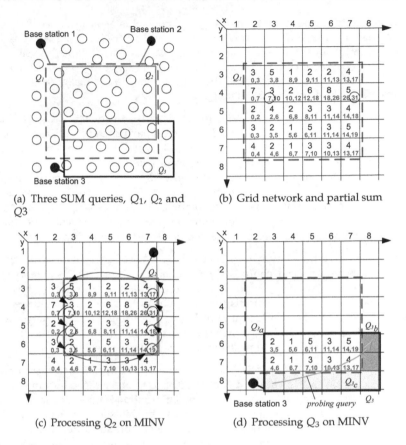

(a) Three SUM queries, Q_1, Q_2 and Q_3

(b) Grid network and partial sum

(c) Processing Q_2 on MINV

(d) Processing Q_3 on MINV

Fig. 3. Materialized in-network view

On the other hand, two major issues are faced in the development of MINV. The first and most critical issue is the presentation and placement of queried results. This directly affects the usability of cached data for any subsequent query. Another issue is about how a query can be processed if its answer is partially or fully available from the cache.

In MINV, we consider a sensed area structured into a grid as shown in Figure 3(b), as opposed to building any ad hoc routing structure that favors queries issued by some base stations at query time. Within every grid cell denoted by $cell(x,y)$, sensor nodes form a cluster and one of the sensor nodes is elected as a cluster head. Upon receiving a query, the cluster head collects sensed readings from all cluster members. Based on this setting, we can treat a sensor network as a grid of cluster heads. To answer aggregation queries, we assume parallel itinerary-based data aggregation as discussed in the previous section. Here, cluster heads serve as Q-nodes, forwarding queries and computing intermediate results. Additional to query processing, cluster heads cache every intermediate query result it receives and that it send. For grid cell $cell(x,y)$, we denote the received intermediate query result as $init(x,y)$ and the sent intermediate query result as $final(x,y)$. As shown in Figure 3(b), intermediate results derived and maintained for a SUM query (called *partial sum*) are accumulated and cached in cluster heads within queried regions. In the figure, cluster head at $cell(3,4)$ maintains an initial partial sum (i.e., $init(3,4)$) and a final partial sum (i.e., $final(3,4)$) as 7 and 10, respectively, while its local reading is 3. Based on cached partial sums, the sum of sensed readings in all cell between $cell(x,y)$ and $cell(x',y)$ in the same row y can be determined as $final(x',y) - init(x,y)$. As in the figure, the sum of sensed readings of sensor nodes from $cell(3,4)$ through $cell(7,4)$ can be calculated as $31 - 7 = 24$.

To answer another SUM query Q_2 whose region is fully covered by Q_1's, Q_2 can simply traverse the border of its query region to collect cached partial sums. In Figure 3(c), Q_2 sums up $init(3,3)$, $init(4,3)$, $init(5,3)$ and $init(6,3)$, i.e., $3 + 7 + 2 + 3 = 15$, from the left side of its query region. Thereafter, it calculates the sum of $final(6,7)$, $final(5,3)$, $final(4,3)$ and $final(3,7)$, i.e., $19 + 18 + 31 + 17 = 85$ from the right side of the region, and subtracts 15 from it. Now the final sum is 70. Notice that only cluster heads on the border of a query region are accessed for cached partial sums and participate in query passing. By using the cache, messages between cluster heads and their members are saved. Besides, some internal grid cells inside a given query region are not accessed at all, further reducing energy costs.

Some queries may have their query regions partially covered by previous queries. In these cases, those queries need to be decomposed into subqueries, which each subquery covers one disjointed subregion. The final query result is then computed by aggregating those subquery results. For instance, Q_3's region is partially covered by Q_1's. Thus, it is partitioned into three subqueries Q_{3a}, Q_{3b} and Q_{3c} as illustrated in Figure 3(d). While Q_{3a} is totally answered by the cached partial sums, Q_{3b} and Q_{3c} are performed as separate SUM queries. The answer of Q_3 is then obtained by adding the sums from these subqueries.

Thus far, cache information has been implicitly assumed to be available to every base stations in the above discussion. In fact, it is not energy efficient to make cache information available everywhere. In MINV, we consider that the cache information is only maintained with initial and final intermediate results in queried grid cells. In this setting, cache discovery is an issue to consider. To determine whether a cache is available for a query, we introduced a probing stage in every query evaluation as illustrated in Figure 3(d). The main idea of this probing stage is described as follows. When a query reaches the (nearest) corner of a query region, it traverses to the diagonally opposite corner and checks if available cache is present in the traversed cells on a diagonal line. If no cache is discovered, it means two possible implications: (i) no cache is available inside the query region, or (ii) a cache if exists has a small overlapped area with the query region, so that it is considered to be not useful to the query. If no cache is used, the query is executed directly from the farthest corner. Otherwise, the query is transformed into subqueries accessing the cache and deriving aggregated reading values in remaining divided areas. Notice that this additional probing stage introduces a little extra communication cost, compared to evaluating queries directly, which usually derives query results at the farthest corners of query regions and sends the results from there back to base stations. Besides, for some cases like entire query regions fully covered by a cache (e.g., Q_2 as discussed above), probe stages can be omitted.

5. In-network contour map computation

As discussed in the previous two sections, data aggregation was used to compute a single aggregated value representing the measurements for an entire sensed area or a query region. For a large sensed area, certain measurements recorded by sensor nodes, e.g., temperature, wind speed, etc., should continuously change over the area. Data aggregation cannot effectively represent such spatially varied measurements. Thus, some other presentations, e.g., histogram, contour map, etc., should be used instead. Among those, contour maps are often used to present the approximate spatial distributions of measurements. On a contour map as illustrated in Figure 4(a), an area is divided into regions by some curves called *contour lines* and every contour line is labeled with one value. Thus, on a contour map, all measurements on a contour line labeled with v are equal to v, whereas measurements at some points not on any contour lines can be determined through interpolation according to their straight-line distances to adjacent contour lines.

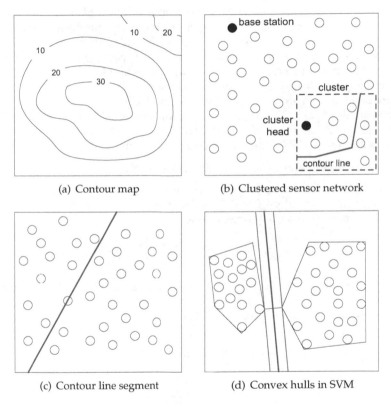

(a) Contour map   (b) Clustered sensor network

(c) Contour line segment   (d) Convex hulls in SVM

Fig. 4. Contour map computation

Very recently, the research of contour map computation in sensor networks has started to receive attention Liu & Li (2007); Meng et al. (n.d.); Xue et al. (2006). An earlier work Xue et al. (2006) was proposed to construct a contour map as a grid, in which each grid cell carries an aggregated single value. This grid presentation can facilitate recognition and matching spatial patterns of measurements with respect to some predefined patterns for event detection and phenomenon tracking. However, the grid presentation cannot provide very precise contour maps and it may incur a large communication cost to convey individual grid cell values, especially when grids of very fine granularity are used.

Motivated by the importance of contour map in sensor networks, we have developed a Contour Map Engine (CME) to compute contour map in sensor networks Xu et al. (2008). More precisely, CME computes contour lines, which can be represented by the coefficients of certain curve/line equations, and thus are small to transmit. In a sensor network, every small area is assumed to be monitored by a cluster of sensor nodes as shown in Figure 4(b). Periodically, a cluster head collects sensed readings from all sensor nodes. Based on their spatial locations and reported sensed readings, the cluster head determines a contour line segment for the area and sends it to a base station. Finally, the base station connects all received contour line segments and constructs a contour map.

Logically, a contour line with respect to a given v_c divides a given area into subareas on its two sides as in Figure 4(c). On one side, all sensor nodes provides reading values not greater than v_c, whereas all other sensor nodes on another side have their readings not smaller than v_c. Here, some sensor nodes reporting their sensed readings of v_c may be distributed around the contour line. Further, given the reading values and locations of individual sensor nodes, partitioning an area by a contour line segment is somewhat equivalent to a binary classification problem. In light of this, the design of CME uses support vector machine (SVM) Christianini & Shawe-Taylor (2000), a commonly used data mining technique, to determines contour line segments. In a cluster of sensor nodes N', each sensor node n ($\in N'$) provides its location x_n and its classified value y_n, which can be either -1 or $+1$, according to its own sensed reading v_n and the contour line value v_c. Here, $y_n = \begin{cases} +1 & v_n \geq v_c \\ -1 & v_n < v_c \end{cases}$. Next, we define the classification boundary (i.e., the contour line segment) as a hyperplane by a pair of coefficients (w, b) such that $w^T x + b = 0$. Based on this, we can estimate an expected $\hat{y}$ for any location x, which may not have any sensor node as

$$\hat{y} = sgn(w^T x + b) = \begin{cases} +1 & w^T x + b \geq 0 \\ -1 & w^T x + b < 0 \end{cases}$$

Now, the classification boundary in SVM is derived to maximize the margin between the convex hull of the two sets, such that classification error for unknown locations can be minimized as depicted in Figure 4(d). The distance between any location x and the classification boundary is $\frac{|w^T x|}{||w||}$. The optimal classification boundary is derived by maximizing the margin, which can be written with Largrange multipliers α_n below:

$$\max_{\alpha} W(\alpha) = \sum_{n \in N'} \alpha_n - \frac{1}{2} \sum_{n \in N'} \sum_{m \in N'} \alpha_n \alpha_m y_n y_m x_n^T x_m$$

subject to $\alpha_n > 0$ and $\sum_{n \in N'} \alpha_n y_n = 0$. Finally, $\max_{\alpha} W(\alpha)$ can be solved by traditional quadratic optimization.

Thus far, our discussion has assumed a single linear contour line segment formed. To handle non-linear classification, our CME utilizes space transformation to divide sensor nodes in a sub-cluster, according to some sample training data. Then, contour line segments are derived from individual sub-clusters. Interested readers can be refer the details in Xu et al. (2008). Some other recent works (e.g., Zhou et al. (2009)) have been presented in the literature to improve the precision of contour line segments by using more sophnicated techniques.

6. In-network probabilistic data aggregation

Sensor reading values are inherently noisy and somewhat uncertain, because of possible inaccurate sensing, environmental noise, hardware defeats, etc., Thus, data uncertainty is another important issue in sensor data analysis. In the literature, uncertain data management has been extensively studied and various models are developed to provide the semantics of underlying data and queries Faradjian et al. (2002); Prabhakar & Cheng (2009). However, existing works adopts centralized approaches Faradjian et al. (2002); Prabhakar & Cheng (2009) that, however, is energy inefficient as already discussed. In-network uncertain data aggregation appears to be new research direction.

Very recently, we have started to investigate a variety of in-network data aggregation techniques for some common aggregation queries. In the following, we discuss one of our recent works on probabilistic minimum value query (PMVQ) Ye, Lee, Lee, Liu & Chen (to appear). A probability minimum value query searches for possible minimum sensed reading value(s).

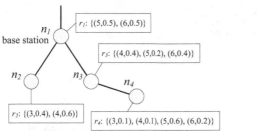

Value v	$Pr[v_{min} = v]$
3	0.46
4	0.54
5	0.00
6	0.00

(a) An example sensor network of four sensor nodes

(b)   Minimum   value probability

Fig. 5. Example sensor network and minimum value probability

Figure 5(a) shows an example sensor network of four sensor nodes. Each sensor node n_i maintains a probabilistic sensed reading r_i, i.e., a set of possible values $\{v_{i,1}, \cdots v_{i,|r_i|}\}$. Each value $v_{i,k}$ is associated with a non-zero probability $p_{i,k}$ being a real sensed reading value. The sum of all $p_{i,k}$ ($1 \leq k \leq |r_i|$) equals 1. The sensed reading r_i of each example sensor node n_i is shown next to the node. For n_1, the actual sensed reading value may be either 5 with a probability of 0.5 or 6 with the same probability. Since every sensed reading has different possible values, it is apparently not trivial to say that 3, which is the smallest possible value among all, is the minimum since it may not actually exist. On the other hand, 4 can be the true minimum when 3 is not real. As such, more than one value can be the minimum value, simultaneously. Thus, the minimum value probability for v being the minimum v_{min} among all possible sensed reading values, denoted by $Pr[v_{min} = v]$, is introduced and defined as below:

$$Pr[v_{min} = v] = \prod_{n_i \in N} Pr[r_i \geq v] - \prod_{n_i \in N} Pr[r_i > v]$$

In our example, $Pr[v_{min} = 3]$ is equal to $(1 \cdot 1 \cdot 1 \cdot 1) - (1 \cdot 0.6 \cdot 1 \cdot 0.9) = 0.46$, $Pr[v_{min} = 4]$ is equal to $(1 \cdot 0.6 \cdot 1 \cdot 0.9) - (1 \cdot 0 \cdot 0.6 \cdot 0.8) = 0.54$, and both $Pr[v_{min} = 5]$ and $Pr[v_{min} = 6]$ are 0, as listed in Figure 5(b). Hence, the minimum value query result include 3 and 4 and their minimum value probabilities are greater than 0.

To evaluate PMVQ in sensor networks, we have devised two algorithms, namely, *Minimum Value Screening (MVS) algorithm* and *Minimum Value Aggregation (MVA) algorithm*. Both of the algorithms evaluate PMVQs in sensor networks organized as routing trees. We describe them in the following.

MVS Algorithm. Suppose that there are two probabilistic sensed readings r_i and r_j from two sensor nodes n_i and n_j, where $r_i = \{v_{i,1}, \cdots v_{i,|r_i|}\}$ and $r_j = \{v_{j,1}, \cdots v_{j,|r_i|}\}$. A value v_j ($\in r_j$) is certainly not the minimum if r_i has all its values smaller than it, i.e., $\forall_{v_i \in r_i} v_i < v_j$. Then, v_j can be safely discarded. Based on this idea, we introduced a notion called MiniMax. Among sensed readings from a subset of sensor nodes N', a MiniMax denoted by MiniMax(N') represents the largest possible value, formally, MiniMax$(N') = \min_{n_i \in N'} \left\{ \max_{v_i \in r_i} \{v_i\} \right\}$.

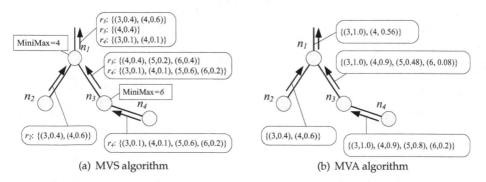

(a) MVS algorithm   (b) MVA algorithm

Fig. 6. MVS and MVA algorithms

This MiniMax notion is used to screen out those values that should not be minimum values. We use Figure 6(a) to illustrate how MiniMax is determined and used by MVS algorithm to eliminate some values and their probabilities from being propagated in a routing tree. First, n_4 sends its sending reading values to n_3, which in turn deduces MiniMax$(\{n_3, n_4\})$, i.e., 6. Thus, n_3 propagates all its and n_4's sensed reading values to n_1. On the other hand, n_2 submits its sensed reading values to n_1. Now, n_1, i.e., the base station, determines MiniMax$(\{n_1, n_2, n_3, n_4\})$, which equals 4. Thus, only n_2's $\{(3, 0.4), (4, 0.6)\}$, n_3's $\{(4, 0.4\}$ and n_4's $\{(3, 0.1), (4, 0.4)\}$ are further propagated to the connected terminal. Later, it determines the final result values according to their minimum value probabilities.

MVA Algorithm. MVA algorithm computes $Pr[v_{min} = v]$ for each candidate value v incrementally during data propagation since computation of $Pr[v_{min} = v]$ is decomposable. Recall that $Pr[v_{min} = v]$ is computed based on two terms, i.e., $\prod_{n_i \in N} Pr[r_i \geq v]$ and $\prod_{n_i \in N} Pr[r_i > v]$. These two terms can be factorized when N is divided into x disjointed subsets, i.e., $N_1, N_2, \cdots N_x$ as follows:

$$\prod_{n_i \in N} Pr[r_i \geq v] = \prod_{i \in [1,x]} \prod_{n_i \in N_i} Pr[r_i \geq v], \qquad \prod_{n_i \in N} Pr[r_i > v] = \prod_{i \in [1,x]} \prod_{n_i \in N_i} Pr[r_i > v]$$

Based on this, in any subtree covering some sensor nodes N_i, the root can calculate $\prod_{n_i \in N_i} Pr[r_i \geq v]$ and $\prod_{n_i \in N_i} Pr[r_i > v]$ for every value v. Then, only the value and these two terms are sent to its parent instead of all individual sensed reading values as needed by MVS algorithm.

Further, due to the fact that $Pr[v_{min} = v]$ should be zero whenever $\prod_{n_i \in N_i} Pr[r_i \geq v] = \prod_{n_i \in N_i} Pr[r_i > v]$ for any non-empty N_i, it is safe to omit value v from being propagated. In addition, for integer sensed reading values, $\prod_{n_i \in N_i} Pr[v_{min} > v]$ should be equal to $\prod_{n_i \in N_i} Pr[v_{min} \geq v+1]$. Therefore, either $\prod_{n_i \in N_i} Pr[v_{min} > v]$ or $\prod_{n_i \in N_i} Pr[v_{min} \geq v+1]$ can be sent to a parent node and the omitted probabilities can be deduced by the parent node.

Figure 6(b) illustrates MVA algorithm. First, n_4 sends each of its value v and $Pr[v_{min} \geq v]$, i.e., $(3, 1.0)$, $(4, 0.9)$, $(5, 0.8)$, $(6, 0.2)$ to n_3. Similarly, n_2 sends $(3, 1.0)$ and $(4, 0.6)$ to n_1. Then, n_3 calculates $Pr[v_{min} \geq v]$ for all its know values, i.e., 3, 4, 5 and 6. Next, n_3 forwards $(3, 1.0)$, $(4, 0.9)$, $(5, 0.48)$ and $(6, 0.8)$ to n_1. Further, n_1 computes $Pr[v_{min} = v]$ as n_3. However, $Pr[v_{min} = 5]$ and $Pr[v_{min} = 6]$ are both 0, so 5 and 6 are filtered out. At last, n_1's $Pr[v_{min} = 3]$ and $Pr[v_{min} = 4]$ are determined and they are equal to zero; and both 3 and 4 are the query result.

Compared with MVS algorithm, MVA algorithm considerably saves communication costs and battery energy. Through detailed cost analysis and simulation experiments as in Ye, Lee, Lee, Liu & Chen (to appear), MVA algorithm provides costs linear to the number of sensor nodes, while MVS incurs significantly large communication costs with respect to the increased number of sensor nodes.

In addition to probabilistic minimum query, we have also investigated other probabilistic queries in sensor networks, e.g., probabilistic minimum node query (PMNQ) Ye, Lee, Lee, Liu & Chen (to appear) that searches for sensor nodes that provide probabilistic minimum values and probabilistic top-k value query that search for k smallest (or largest) values Ye, Lee, Lee & Liu (to appear).

7. Summary and future directions

Wireless sensor networks are important tools for many fields and applications. Meanwhile, in sensor networks, data acquisition that collects data from individual sensor nodes for analysis is one of the essential activities. However, because of scarce sensor node battery energy, energy efficiency becomes a critical issue for the length of sensor network operational life. Over those years, many research works have studied various in-network query processing as one of the remedies to precious precious sensor node energy. By in-network query processing, queries are disseminated and processed by sensor nodes and a small volume of (derived) data is collected and transmitted rather than raw sensed readings over costly wireless communication. Subject to the supported types of queries and potential optimizations, a variety of in-network query processing techniques have been investigated and reported in the literature.

This chapter is devoted to review representative works in in-network data aggregation, data caching, contour map computation and probabilistic data aggregation. With respect to those areas, we also discussed our recent research results, namely, itinerary-based data aggregation, materialized in-network view, contour mapping engine and probabilistic minimum value search. Itinerary-based data aggregation navigates a query among sensor nodes in a queried region for an aggregated value. Compared with infrastructure-based approaches, it incurs fewer rounds of messages and can easily deal with sensor node failure in the course of query processing. To boost the performance of multi-queries issued from different base stations, materialized in-network views provide partial results for previous queries to subsequent aggregation queries. It is different from existing works that cache sensed readings independently and that cannot directly support data aggregation. Contour mapping engine adopts data mining techniques to determine contour line segments in sensor networks, whereas some other works relies on centralized processing or provide less accurate contour maps. Last but not least, probabilistic minimum value search is the initial research result on uncertain sensed data aggregation. As sensed reading values are mostly imprecise, handling and querying probabilistic sensor data is currently an important on-going research direction.

In addition, recent research studies have shown uneven energy consumption of sensor nodes that sensor nodes in some hotspot regions have more energy consumed than others Perillo et al. (2005). Such hotspot problems are currently studied from the networking side. Besides, heterogeneous sensor nodes are going to be very common in sensor networks. Thus, we anticipate that future in-network query processing techniques should be able to handle uneven energy consumption and to make use of super sensor nodes, while many existing works mainly presume homogeneous sensor nodes and consider even energy consumption.

8. References

Biswas, R., Thrun, S. & Guibas, L. J. (2004). A Probabilistic Approach to Inference with Limited Information in Sensor Networks, *Proceedings of the Third International Symposium on Information Processing in Sensor Networks (IPSN), Berkeley, CA, Apr 26-27*, pp. 269–276.

Bojkovic, Z. & Bakmaz, B. (2008). A Survey on Wireless Sensor Networks Deployment, *WSEAS Transactions on Communications* 7(12).

Christianini, N. & Shawe-Taylor, J. (2000). *An Introduction to Support Vector Machines and Other Kernel-Based Learning Methods*, Cambridge University Press.

Doherty, L. & Pister, K. S. J. (2004). Scattered Data Selection for Dense Sensor Networks, *Proceedings of the Third International Symposium on Information Processing in Sensor Networks (IPSN), Berkeley, CA, Apr 26-27*, pp. 369–378.

Faradjian, A., Gehrke, J. & Bonnet, P. (2002). GADT: A Probability Space ADT for Representing and Querying the Physical World, *Proceedings of the 18th IEEE International Conference on Data Engineering (ICDE), San Jose, CA, Feb 26 - Mar 1*, pp. 201–211.

Faulkner, M., Olson, M., Chandy, R., Krause, J., Chandy, K. M. & Krause, A. (2011). The Next Big One: Detecting Earthquakes and Other Rare Events from Community-Based Sensors, *Proceedings of the 10th International Conference on Information Processing in Sensor Networks (IPSN), Chicago, IL, Apr 12-14*, pp. 13–24.

Ganesan, D., Estrin, D. & Heidemann, J. S. (2003). Dimensions: Why Do We Need a New Data Handling Architecture for Sensor Networks?, *Computer Communication Review* 33(1): 143–148.

Group, T. W. (n.d.). TinyOS, http://www.tinyos.net/.

Hellerstein, J. M., Hong, W., Madden, S. & Stanek, K. (2003). Beyond Average: Toward Sophisticated Sensing with Queries, *Proceedings of Information Processing in Sensor Networks, Second International Workshop (IPSN), Palo Alto, CA, Apr 22-23*, pp. 63–79.

Huang, Z., Wang, L., Yi, K. & Liu, Y. (2011). Sampling Based Algorithms for Quantile Computation in Sensor Networks, *Proceedings of the ACM SIGMOD International Conference on Management of Data (SIGMOD), Athens, Greece, Jun 12-16*, pp. 745–756.

Invanov, A. O. & Tuzhilin, A. A. (1994). *Minimal Networks: The Steiner Problem and Its Generalizations*, CRC Press.

Lee, K. C. K., Zheng, B., Lee, W.-C. & Winter, J. (2007). Materialized In-Network View for Spatial Aggregation Queries in Wireless Sensor Network, *ISPRS Journal of Photogrammetry and Remore Sensing* 62: 382–402.

Li, Z., Zhu, Y., Zhu, H. & Li, M. (2011). Compressive Sensing Approach to Urban Traffic Sensing, *Proceedings of IEEE International Conference on Distributed Computing Systems (ICDCS), Minneapolis, MN, Jun 20-24*, pp. 889–898.

Liu, X., Huang, Q. & Zhang, Y. (2004). Combs, Needles, Haystacks: Balancing Push and Pull for Discovery in Large-Scale Sensor Networks, *Proceedings of the 2nd ACM International Conference on Embedded Networked Sensor Systems (SenSys), Baltimore, MD, Nov 3-5*, pp. 122–133.

Liu, Y. & Li, M. (2007). Iso-Map: Energy-Efficient Contour Mapping in Wireless Sensor Networks, *Proceedings of the 27th IEEE International Conference on Distributed Computing Systems (ICDCS), Toronto, Ontario, Canada, Jun 25-29*, p. 36.

Madden, S. & Franklin, M. J. (2002). Fjording the Stream: An Architecture for Queries Over Streaming Sensor Data, *Proceedings of the 18th IEEE International Conference on Data Engineering, San Jose, CA, Feb 26 - Mar 1*, pp. 555–566.

Madden, S., Franklin, M. J., Hellerstein, J. M. & Hong, W. (2002). TAG: A Tiny AGgregation Service for Ad-Hoc Sensor Networks, *Proceedings of The 5th USENIX Symposium on Operating System Design and Implementation (OSDI), Boston, MA, Dec 9-11*.

Manjhi, A., Nath, S. & Gibbons, P. B. (2005). Tributaries and Deltas: Efficient and Robust Aggregation in Sensor Network Streams, *Proceedings of the ACM SIGMOD International Conference on Management of Data (SIGMOD), Baltimore, MD, Jun 14-16*, pp. 287–298.

Meng, X., Nandagopal, T., Li, L. & Lu, S. (n.d.). Contour Maps: Monitoring and Diagnosis in Sensor Networks, 50(15): 2920–2838.

Musaloiu-Elefteri, R., Liang, C.-J. M. & Terzis, A. (2008). Koala: Ultra-Low Power Data Retrieval in Wireless Sensor Networks, *Proceedings of the 7th International Conference on Information Processing in Sensor Networks (IPSN), St. Louis, MO, Apr 22-24*, pp. 421–432.

Perillo, M. A., Cheng, Z. & Heinzelman, W. B. (2005). An Analysis of Strategies for Mitigating the Sensor Network Hot Spot Problem, *Proceedings of the 2nd Annual International Conference on Mobile and Ubiquitous Systems (MobiQuitous), San Diego, Jul 17-21*, pp. 474–478.

Prabh, S. & Abdelzaher, T. F. (2005). Energy-Conserving Data Cache Placement in Sensor Networks, *ACM Transactions on Sensor Networks* 1(2): 178–203.

Prabhakar, S. & Cheng, R. (2009). Data Uncertainty Management in Sensor Networks, *Encyclopedia of Database Systems*, pp. 647–651.

Ratnasamy, S., Karp, B., Yin, L., Yu, F., Estrin, D., Govindan, R. & Shenker, S. (2002). GHT: a Geographic Hash Table for Data-Centric Storage, *Proceedings of the First ACM International Workshop on Wireless Sensor Networks and Applications (WSNA), Atlanta, GA, Sept 28*, pp. 78–87.

Sadagopan, N., Krishnamachari, B. & Helmy, A. (2003). The ACQUIRE Mechanism for Efficient Querying in Sensor Networks, *IEEE International Workshop on Sensor Network Protocols and Applications (SNPA'03), held in conjunction with the IEEE International Conference on Communications (ICC), Anchorage, AL*.

Shakkottai, S. (2004). Asymptotics of Query Strategies over a Sensor Network, *Proceedings of The 23rd IEEE Annual Joint Conference of the IEEE Computer and Communications Societies (INFOCOM), Hong Kong, China, Mar 7-11.*

Silberstein, A., Braynard, R., Ellis, C. S., Munagala, K. & Yang, J. (2006). A Sampling-Based Approach to Optimizing Top-k Queries in Sensor Networks, *Proceedings of the 22nd International Conference on Data Engineering (ICDE), Atlanta, GA, Apr 3-8,* p. 68.

Vasilescu, I., Kotay, K., Rus, D., Dunbabin, M. & Corke, P. I. (2005). Data Collection, Storage, and Retrieval with an Underwater Sensor Network, *Proceedings of the 3rd ACM International Conference on Embedded Networked Sensor Systems (SenSys), San Diego, CA, Nov 2-4,* pp. 154–165.

Wu, S.-H., Chuang, K.-T., Chen, C.-M. & Chen, M.-S. (2007). DIKNN: An Itinerary-based KNN Query Processing Algorithm for Mobile Sensor Networks, *Proceedings of the 23rd IEEE International Conference on Data Engineering (ICDE), Istanbul, Turkey, Apr 15-20,* pp. 456–465.

Xu, Y., , Lee, W.-C. & Mitchell, G. (2008). CME: A Contour Mapping Engine in Wireless Sensor Networks, *The 28th International Conferences on Distributed Computing Systems (ICDCS), Beijing, China, Jun 17-20,* pp. 133–140.

Xu, Y., Lee, W.-C. & Xu, J. (2007). Analysis of A Loss-Resilient Proactive Data Transmission Protocol in Wireless Sensor Networks, *Proceedings of 26th IEEE International Conference on Computer Communications, Joint Conference of the IEEE Computer and Communications Societies (INFOCOMM), Anchorage, AL, May 6-12,* pp. 1712–1720.

Xu, Y., Lee, W.-C., Xu, J. & Mitchell, G. (2006). Processing Window Queries in Wireless Sensor Networks, *Proceedings of the 22nd International Conference on Data Engineering (ICDE), Atlanta, GA, Apr 3-8,* p. 70.

Xue, W., Luo, Q., Chen, L. & Liu, Y. (2006). Contour Map Matching for Event Detection in Sensor Networks, *Proceedings of the ACM SIGMOD International Conference on Management of Data, Chicago, IL, Jun 27-29,* pp. 145–156.

Yao, Y. & Gehrke, J. (2003). Query Processing in Sensor Networks, *Online Proceedings of The First Biennial Conference on Innovative Data Systems Research (CIDR), Asilomar, CA, Jan 5-8.*

Ye, M., Lee, K. C. K., Lee, W.-C., Liu, X. & Chen, M. C. (to appear). Querying Uncertain Minimum in Wireless Sensor Networks, *IEEE Transactions on Knowledge and Data Engineering* .

Ye, M., Lee, W.-C., Lee, D. L. & Liu, X. (to appear). Distributed Processing of Probabilistic Top-k Queries in Wireless Sensor Networks, *IEEE Transactions on Knowledge and Data Engineering* .

Zhang, W., Cao, G. & Porta, T. L. (2007). Data Dissemination with Ring-Based Index for Wireless Sensor Networks, *IEEE Transactions on Mobile Computing* 6(7): 832–847.

Zhou, Y., Xiong, J., Lyu, M. R., Liu, J. & Ng, K.-W. (2009). Energy-Efficient On-Demand Active Contour Service for Sensor Networks, *Proceedings of IEEE 6th International Conference on Mobile Adhoc and Sensor Systems (MASS), Macau, China, Oct 12-15,* pp. 383–392.

Zurich, T. W. R. G. . E. (n.d.). The Sensor Network Museum, http://www.snm.ethz.ch/Main/HomePage.

9

Differential Absorption Microwave Radar Measurements for Remote Sensing of Barometric Pressure

Roland Lawrence[1], Bin Lin[2], Steve Harrah[2] and Qilong Min[3]
[1]*Old Dominion University*
[2]*NASA Langley Research Center*
[3]*SUNY at Albany*
USA

1. Introduction

1.1 Overview

As coastal regions around the world continue to grow and develop, the threat to these communities from tropical cyclones also increases. The predicted sea level rise over the next decades will certainly add to these risks. Developed low-lying coastal regions are already of major concern to emergency management professionals. While hurricane forecasting is available, improved predictions of storm intensity and track are needed to allow the time to prepare and evacuate larger cities. The predictions and forecasts of the intensity and track of tropical storms by regional numerical weather models can be improved with the addition of large spatial coverage and frequent sampling of sea surface barometry. These data are critically needed for use in models.

This chapter will present recent advances in the development of a microwave radar instrument technique to remotely sense barometric pressure over the ocean and may provide the large-scale sea surface barometric pressure data needed to substantially improve the tropical storm forecasts. The chapter will include a brief introduction, a discussion of the applications of remote sensing of sea surface barometric pressure, a discussion of the theoretical basis for the differential absorption radar concept, the results of laboratory and flight testing using a prototype radar, and a detailed discussion of the performance challenges and requirements of an operational instrument.

1.2 Background

Surface air pressure is one of the most important atmospheric parameters that are regularly measured at ground based surface meteorological stations. Over oceans, sea surface air barometric pressures are usually measured by limited numbers of in-situ observations conducted by buoy stations and oil platforms. The spatial coverage of the observations of this dynamically critical parameter for use by weather forecasters is very poor. For example, along the east coast of the United States and Gulf of Mexico, only about 40 buoys are available under the NOAA Ocean Observing System (NOOS) of the NOAA National Data Buoy Center (NDBC; http://www.ndbc.noaa.gov/). The tropical atmosphere ocean (TAO) program only has 10 sites from which the barometric pressure is measured. For severe weather conditions, such as tropical storms and hurricanes, these NOOS and TAO buoy systems usually cannot provide spatially desirable in-situ measurements due to either the lack of buoy stations along the actual track of the storm or malfunctions of buoys caused by the severe weather itself.

Under tropical cyclone conditions, including tropical depression, tropical storm, hurricane, and super-typhoon cases, the surface barometric pressure is one of the most important meteorological parameters in the prediction and forecast of the intensity and track of tropical storms and hurricanes. The central air pressure at sea level of tropical cyclones is the most commonly used indicator for hurricane intensity. The classification of tropical storms and hurricanes on the Saffir-Simpson Hurricane Scale (SSHS) is based on the maximum sustained surface wind speed that is a direct result of the interaction between the central air pressure and the pressure fields surrounding tropical storms. Because intensity predictions and landfall forecasts heavily rely upon them, measurements of the central pressure of tropical storms are extremely important. The only method currently available for use is a manned aircraft dropsonde technique. The problem with the dropsonde technique is that each dropsonde supplies only one spatial point measurement at one instant of interest during the passage of the storm. This limits data to the number of dropsondes used and their spatial distribution and thereby leaves most of the storm area unmeasured. Furthermore, dropsondes are difficult to precisely position and cannot be reused. Figure 1 shows the current capability for sea surface barometric measurements; all of them are in situ observations.

To improve predictions and forecasts of the intensity and track of tropical storms, large spatial coverage and frequent sampling of sea surface barometry are critically needed for use in numerical weather models. These needed measurements of sea surface barometric pressure cannot be realized by in-situ buoy and aircraft dropsonde techniques. One approach that may provide barometry in large spatial and temporal scales over oceans is the use of remote sensing techniques including those on board manned aircraft, unmanned aerial vehicles (UAVs), and satellite platforms.

During the last two decades, the development of remote sensing methods, especially airborne and satellite techniques, for large and global scale sea surface pressure measurements significantly lagged methods for other important meteorological parameters,

Fig. 1. Drift Buoy (left), Moored Buoy (middle), and Dropsonde (right).

such as temperature and humidity. There have been suggestions for using satellite oxygen A-band methods, both passive and active, to measure pressure (Barton & Scott, 1986; Korb & Weng, 1982; Singer, 1968; Wu, 1985; and references therein). The active instruments rely on the operation of complicated, highly-stable laser systems on a space platform and are thus technically difficult. Passive methods are restricted to daytime measurements and areas of low cloud cover (Barton & Scott, 1986). Although substantial research efforts have been underway, there are no realizations of remote sensing measurements for atmospheric surface pressure presently available.

This chapter will describe the development of an active microwave radar working at moderate to strong O_2 absorption bands in the frequency range of 50~56 GHz for surface barometric pressure remote sensing, especially over oceans. The sensor concept and flight testing of a proof-of-concept O_2-band radar system for sea surface air pressure remote sensing will also be discussed. At these radar wavelengths, the reflection of radar echoes from water surfaces is strongly attenuated by atmospheric column O_2 amounts. Because of the uniform mixture of O_2 gases within the atmosphere, the atmospheric column O_2 amounts are proportional to atmospheric path lengths and atmospheric column air amounts, thus, to surface barometric pressures. Historically, (Flower & Peckham, 1978) studied the possibility of a microwave pressure sounder using active microwave techniques. A total of six channels covering frequencies from ~25GHz to ~75GHz were considered. A major challenge in this approach is the wide spectral region and the significant additional dependence of radar signals on microwave absorption from liquid water (LW) clouds and atmospheric water vapor (WV) over this range of frequencies. Atmospheric and cloud water temperatures also have different effects on the absorptions at different wavelengths (Lin et al., 1998a, 1998b, 2001). The complexity in matching footprints and obtaining accurate surface reflectivities of the six different wavelength channels makes their system problematic (Barton & Scott, 1986). Recently, (Lin & Hu, 2005) have considered a different technique that uses a dual-frequency, O_2-band radar to overcome the technical obstacles. They have outlined the characteristics of the novel radar system, and simulated the system performance. The technique uses dual wavelength channels with similar water vapor and liquid water absorption characteristics, as well as similar footprints and sea surface reflectivities, because of the closely spaced spectra. The microwave absorption effects due to LW and WV and the influences of sea surface reflection should be effectively removed by use of the ratio of reflected radar signals of the two channels. Simulated results (Lin & Hu, 2005) suggest that the accuracy of instantaneous surface air pressure estimations from the echo ratio could reach 4 – 7 millibars (mb). With multiple pressure measurements over less than ~1km² sea surface spots from the radar echoes, the pressure estimates could be significantly reduced to a few millibars, which is close to the accuracy of in situ measurements and very useful for tropical storm and large scale operational weather modeling and forecasting over oceans.

2. Sea surface barometric pressure measurements for hurricane forecasts

One of the proposed applications of the Differential Absorption Barometric Radar, hereafter called DiBAR, is to improve weather forecasts and predictions, especially for tropical storms. To address the usefulness of sea surface barometric measurements from DiBAR, we use weather prediction models to simulate predicted hurricane intensities and tracks. Predicted results with sea surface air pressure data incorporated are compared with those without the pressure measurements. These surface pressures were obtained from later analysis of in-situ measurements and the assimilated data of the actual hurricane events. During these actual hurricane events, these sea surface pressure data were not available a priori for modeling and prediction. Quantitative potential improvements in the forecasts and predictions of studied hurricane cases are evaluated. We emphasize that the sea surface air pressure data injected into weather prediction models are not exactly the same as those from later analysis of in-situ measurements and the assimilated data of the actual hurricane events. Some uncertainties exist in the injected pressure data in our simulations to reflect potential DiBAR remote sensing errors, according to our current understanding of DiBAR systems and retrieval uncertainties. This section provides a brief description of the weather forecast model used to simulate the impact of pressure data consistent with our instrument concept, as well as, the results of our study to simulate the improved track and intensity predictions that result from the inclusion of the simulated DiBAR pressure data.

2.1 Weather forecast model description

The numerical weather forecast model used in this study is the Advanced Regional Prediction System (ARPS) developed by the Center for Analysis and Prediction of Storms (CAPS) of the University of Oklahoma and adopted by NASA Langley Research Center (Wang et al., 2001; Xue et al., 2003; Wang & Minnis, 2003). The forward prediction component of the ARPS is a three-dimensional, non-hydrostatic compressible model in a terrain-following coordinate system. The model includes a set of equations for momentum, continuity, potential temperature, water vapor, and turbulence kinetic energy (TKE). It also includes five conservation equations for hydrometeor species: cloud water (small cloud liquid droplets), cloud ice (small ice crystals), rain, snow, and hail (Tao & Simpson 1993). The cloud water and cloud ice move with the air, whereas the rain, snow, and hail fall with their terminal velocity. It has multiple-nested capability to cover the cloud-scale domain and mesoscale domain at the same time. The model employs advanced numerical techniques (e.g., a flux-corrected transport advection scheme, a positive definite advection scheme, and the split-time step). The most unique physical processes included in the model system are a scheme of Kessler-type warm-rain formation and 3-type ice (ice, snow, and hail) microphysics; a soil-vegetation land-surface model; a 1.5-order TKE-based non-local planetary boundary layer parameterization scheme; a cloud-radiation interaction atmospheric radiative transfer scheme; and some cumulus parameterization schemes used for coarse grid-size. Furthermore, a sophisticated long- and short-wave cloud-radiation interaction package (Chou, 1990, 1992; Chou & Suarez, 1994) has been applied to the ARPS model. The ARPS can provide more physically realistic 4D cloud information in very-high-resolution of spatial (cloud processes) and temporal (minutes) scales (Figure. 2).

The ARPS model was run in a horizontal domain of 4800 km, east-west and 4000 km, south-north, and a vertical domain of 25 km. The horizontal grid spacing is 25 km, and the vertical grid space varies from 20 m at the surface to 980 m at the model top. These spatial resolutions are used because they are comparable to those of the models used in the Global Modeling and Assimilation Office, NASA Goddard Space Flight Center. The options for ice microphysics and atmospheric cloud-radiation interactive transfer parameterization were both used in the model. Because of the use of the relatively coarser grid-size of 25 km, the new Kain & Fritsch cumulus parameterization scheme was used together with explicit ice microphysics.

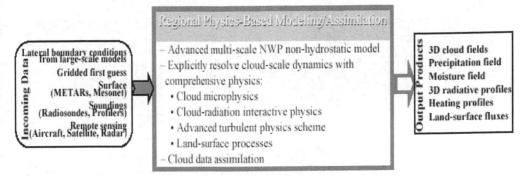

Fig. 2. ARPS: a regional cloud-scale modeling/assimilation system.

2.2 Forecast improvements with the addition of storm central pressure measurements

The analyzed case here is hurricane Ivan (2004). Ivan was a classical, long-lived Cape Verde hurricane that reached Category 5 strength (SSHS) three times and caused considerable damage and loss of life as it passed through the Caribbean Sea. Ivan developed from a large tropical wave accompanied by a surface low-pressure system that moved off the west coast of Africa on 31 August 2004. The development of the system continued and became tropical storm Ivan at 0600 UTC 3 September and a hurricane at 0600 UTC 5 September. After passing Grenada and moving into the southeastern Caribbean Sea, the hurricane's intensity leveled off until 1800 UTC on 8 September when a brief period of rapid intensification ensued. Reconnaissance aircraft data indicated Ivan reached its second peak intensity -- 140 kt and category 5 strength (SSHS) -- just 12 hours later. This was the first of three occasions that Ivan reached the category 5 level.

We choose the forecast period from 0000 UTC 8 Sept. to 0000 UTC 11 Sept. 2004 to examine effects of the central sea surface air pressure on predicting the hurricane track. For the control run (referred as CTL), the model started at 0000 UTC 8 Sep 2004 with the NOAA NCEP Global Forecast System (GFS) analysis fields as the model initial condition. For the central sea level air pressure experiment run (referred as SLP), only the observed central pressure was added to the initialization, using the GFS analysis as the first guess. The lateral boundary conditions for both simulations came from the GFS 6-hour forecasts. The same model physics options were used for the two experiments.

As shown in Figure 3, from run CTL, the hurricane central pressure at the initial time of 0000 UTC 8 Sept 2004 is about 998.7 hPa (obtained from the NOAA/NCEP GFS global large-scale analysis), which is ~15 hPa lower than normal conditions. Although this simulated pressure drop is much smaller than the real hurricane center air pressure depression (see below) and relatively weak for a hurricane, it still could be well captured with our proposed O_2-band radar systems. At 0000 UTC 8 Sept 2004, based on the report of the National Hurricane Center, hurricane Ivan was located at 12.0° N and 62.6° W, and the value of central sea level pressure of the hurricane is actually 950 hPa. This observation-based central pressure estimate was assimilated into the model analysis system. The assimilated initialization field shown in Figure 4 is used as the initial condition in run SLP. The value of the central pressure of the hurricane now is about 951.5 hPa, much closer to the observed

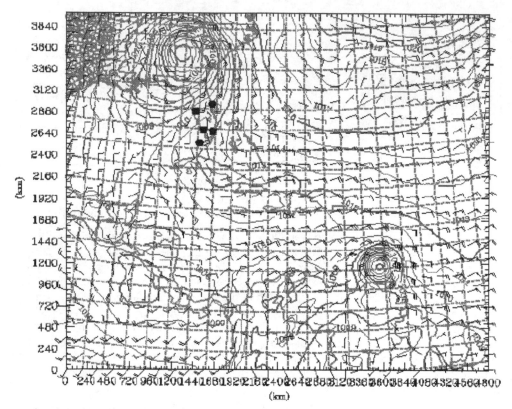

Fig. 3. The sea level air pressure at the initial time of 0000 UTC 8 Sep 2004 for the control run CTL. It is directly interpolated from GFS analysis.

950 hPa and within the error bar of observations. Compared to Figure 3, the change in the initial hurricane center sea level pressure is about 47mb, which significantly improves the predicted hurricane intensity.

The model was integrated for 72 hours at a time step of 15-seconds and used to estimate the storm track. It is not surprising that both of the experiments capture the hurricane track much better than the operational GFS global forecasting (Figure 5). This is mainly because the regional numerical model is non-hydrostatic with explicit cloud/ice-physics parameterizations, cloud-radiation interaction, as well as advanced turbulence schemes, and land-surface interaction. This kind of advanced regional model can better resolve multi-scale atmospheric processes, especially for organized convective cloud systems. A significant improvement in the predicted hurricane track resulted from the use of the observations of the central surface pressure in the initialization of SLP, as shown in Figure 5. The SLP experiment generated a more realistic hurricane track, especially for the first two forecasts. The results of our sensitivity tests suggest that it is possible to make better predictions of hurricane track by using surface pressure observations/measurements within the targeted tropical cyclone region.

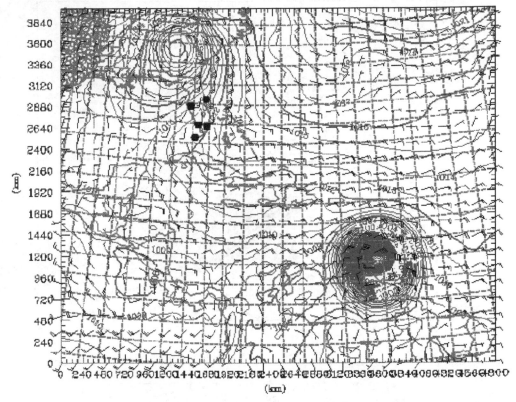

Fig. 4. The sea level pressure at the initial time of 0000 UTC 8 Sep 2004 for the experimental run SLP. The observed central pressure was used for the initialization with GFS analysis as the background.

2.3 Forecast improvements when pressure fields are ingested into model

The results of typical weather predictions for a tropical cyclone, using not only center sea surface air pressures but also large area pressure fields, is shown in Fig. 6 for 1996 hurricane Fran, which occurred from 0000UTC September 3 to 0060 UTC September 6, 1996 (Xiao et al. 2000). Due to the lack of data, the model standard run (control run; CTL curve) started with a location error of about 100km, and gradually deviated from the observed hurricane track (OBS curve) up to about 350km for the predicted landfall site. With pressure data and calculated wind fields as inputs, the assimilations with 54km (A80 curve) and 18km (B80 curve) spatial resolution significantly reduced the errors in predicted storm tracks. Comparing the 3 day forecasts, the high-resolution model (18 km, B80) had a small starting location error of about 10 km that increased to about 100 km at the predicted landfall site, and the low-resolution model (54 km, A80) had a starting error of about 35 km and predicted landfall with a 170 km error. Such greatly improved predictions could make hurricane preparation and evacuation much easier, especially for the high resolution forecast (B80) case.

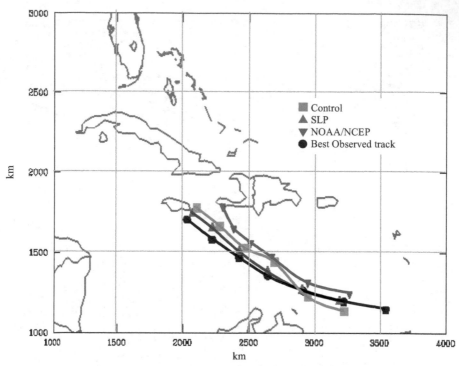

Fig. 5. The predicted hurricane tracks from 0000 UTC 8 Sep 2004 to 0000 UTC 11 Sep 2004.

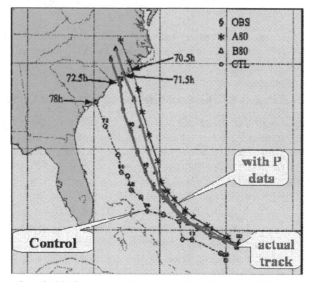

Fig. 6. Predicted tracks of 1996 hurricane Fran by CTL, B80, and A80, along with observations, from 0000 UTC 3 Sep to 0600 UTC 6 Sep. Predicted landing times are also indicated in the figure.

Storm intensity predictions can also be improved with knowledge about the storm center pressure, pressure gradients, and derived wind fields. As expected, the intensity of the B80 prediction is very close to observations at the landfall site (Xiao et al., 2000). The hurricane eye, rain band, and precipitation intensity determined from radar reflectivity simulations (a) and radar observations (b) are very similar (Figure 7). The similarity between these predicted hurricane intensity fields, using pressure fields as one of critical initial conditions, and fields based on observations is remarkable. Unfortunately, there have been no operational, or even experimental, surface air pressure measurements over open oceans

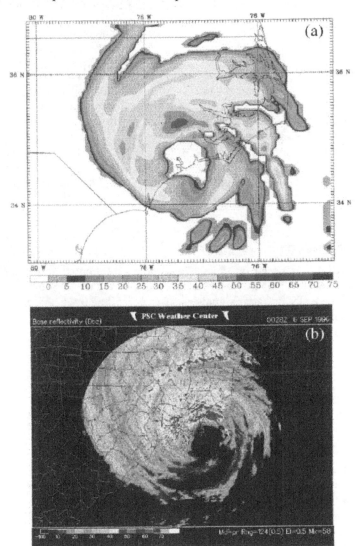

Fig. 7. Radar reflectivity (dBZ) (a) predicted by B80 at 0000 UTC 6 Sep 1996 and (b) captured at Wilmington, NC, at 0028 UTC 6 Sep 1996.

from both in-situ and remote sensing instruments, and thus it remains difficult to predict the tracks and intensities of tropical storms with high accuracies (within 100km landfall site for 3-day forecasts).

The results of the above simulations suggest that tropical storm forecasts of landfall and intensity at landfall may be improved by adding pressure field data consistent with the DiBAR measurement concept. With the pressure measurements of the center and whole field of tropical storms, our simulations using regional weather forecast models show that the prediction of hurricane tracks and intensities can be significantly improved. For the hurricane Fran case, model prediction reduces the landfall site errors from ~350km in the standard prediction to ~100km for 3 day forecasts, which could improve hurricane preparation and evacuation.

An operational airborne instrument could provide unprecedented barometric sampling in terms of spatial coverage and repeat rates. Assuming similar operational flights a DiBAR instrument would be expected to provide data at the same pressure resolution but much higher spatial density. If UAV is used, the cost of providing the needed barometric measurements could be significantly lower than that of current operations using in-situ techniques with the accompanying increase in personnel safety. Future space borne systems may further improve the pressure field sampling, albeit with a more coarse spatial resolution. Furthermore, the availability of these data could result in improved weather forecasts for catastrophic events and could significantly reduce human loss and property damage.

3. Measurement approach

The DiBAR instrument is based on the retrieval of the differential absorption near the O_2 line complex (frequencies: 50–56 GHz). This selection of frequencies provides large changes in absorption for the reflected radar signals as a function of the frequency of the radar due in part to the different atmospheric O_2 attenuation coefficients. In the atmosphere, O_2 is generally uniformly mixed with other gases. The O_2 in the column observed by the radar is proportion to column air mass, the column air mass is proportional to the surface air pressure, and the reflected power measured by the radar can be approximated as (*Lin and Hu* 2005)

$$P_r(f) = \left(\frac{P_T G_t G_r \lambda^2}{(4\pi)^3} \right) \left(\frac{\sigma^0(f)}{r^2} \right) \exp\left(-\frac{2\alpha_o M_o P_o}{g} - 2\alpha_L L - 2\alpha_v V \right) \qquad (1)$$

where the first term in equation (1) includes frequency dependent characteristics of the radar, which must be determined by instrument calibration: P_T is the transmitter power and G represents the transmitter and receiver antenna gain. The second term includes changes in the surface reflectivity, σ^0, over the radar frequency, and the last term represents the atmospheric absorption, where M_0 is the mixing ratio of O_2 to total air and P_o is the surface pressure. Thus, if the frequency response of the radar is well characterized from 50 -56 GHz, and the absorption characteristics due to liquid water and water vapor, and spatial resolution of the radar are similar over this range of frequencies, then the ratio of the radar received powers from two frequencies is then,

$$\frac{P_r(f_1)}{P_r(f_2)} = \left(\frac{C(f_1)}{C(f_2)}\right) \exp\left(-\frac{2(\alpha_o(f_1) - \alpha_o(f_2))M_o P_o}{g}\right) \tag{2}$$

where $C(f)$ is the frequency dependent radar characteristics. Further, if we define the differential absorption index, $Ri(f_1, f_2)$, as the logarithm of the radar return ratio shown in equation (2), then the surface pressure can be written as,

$$P_o = \left(\frac{2(\alpha_o(f_1) - \alpha_o(f_2))M_o}{g}\right)^{-1} \ln\left(\left(\frac{C(f_2)}{C(f_1)}\right)\left(\frac{P_r(f_1)}{P_r(f_2)}\right)\right)$$

$$P_o = \left(\frac{2(\alpha_o(f_1) - \alpha_o(f_2))M_o}{g}\right)^{-1} (Ci(f_1, f_2) + Ri(f_1, f_2)) \tag{3}$$

or defining terms for a linear relationship between Ri and P_o,

$$P_o = C_0(f_1, f_2) + C_1(f_1, f_2)Ri(f_1, f_2) \tag{4}$$

The term $C_0(f_1, f_2)$ includes the instrument residual calibration error. The differential absorption index, $Ri(f_1, f_2)$, is the logarithm of the ratio of the radar return exclusive of the frequency response of the radar. From equation 4, it can be seen that a simple near-linear relationship between surface air pressure and the differential absorption index is expected from the O_2 band radar data. The linear relationship between Ri and the surface pressure was firstly suggested by the results of modeled differential absorption for several frequencies in the range of interest here (*Lin and Hu* 2005). Further, *Lin and Hu* 2005 suggest that the accuracy of instantaneous surface air pressure estimations from the measured Ri could reach 4 – 7 mb. However, the O_2 absorption increases at higher frequencies and the receiver Signal to Noise Ratio (SNR) may limit the retrieval accuracy as this loss increases. For a fixed transmit power the optimum frequencies for the surface pressure measurement will depend on the received power, which depends on the atmospheric loss and surface reflectivity. The flight testing of the DiBAR instrument discussed in Section 4 is intended to measure the atmospheric attenuation as a function of frequency and the differential absorption index $Ri(f_1, f_2)$. These measurements can then be compared to predicted values to assess the measurement approach and the affect of receiver noise on the measurement of barometric pressure.

In addition to the above analysis a multiple layered atmospheric microwave radiative transfer model was also employed to simulate the atmospheric loss. The technique used to simulate the propagation of radar signals within the atmosphere is based on a plane-parallel, multiple layered atmospheric microwave radiative transfer (MWRT) model that has been used to determine cloud liquid/ice water path, column water vapor, precipitation, land surface emissivity and other parameters over land and oceans (Ho et al., 2003; Huang et al., 2005; Lin & Rossow, 1994, 1996, 1997; Lin et al. 1998a, 1998b; Lin & Minnis, 2000). To avoid complexities of microwave scattering by precipitating hydrometeors and surface backscattering, this study deals only with non-rain weather conditions and homogeneous backgrounds (such as sea surface). Thus, transmission and absorption of radar signals within each atmospheric layer are the major radiative transfer processes considered in the model calculations. For the absorption process, this MWRT model carefully accounts for the temperature and pressure dependences of cloud water and atmospheric gas absorptions (Lin et al., 2001). At microwave wavelengths, temperature dependences of gas and water absorptions are significant, and produce some difficulties for MWRT modeling. The several models available to account for gas absorption differ mainly in their treatment of water vapor continuum absorption. The Liebe model i.e., MPM89 was used here (Liebe, 1989). It yields results that differ negligibly from those of the (Rosenkranz, 1998) model at the O_2 bands. Liquid water absorption coefficients were calculated from the empirical water refractive index formulae of (Ray, 1972), which agree well (relative differences < 5%) with those from (Liebe et al., 1991) for T > $-15°$ C. For colder clouds, the uncertainties in the absorption coefficients could be larger by more than 15% (Lin et al., 2001) because of a lack of direct measurements of the refractive index.

Current MWRT model is consistent of 200 constant-thickness layers from surface to 40km. There is virtually no gas absorption above the modeled top-of-atmosphere (TOA) at our considered spectra. The atmospheric profiles of temperature, pressure, humidity and gas amount are obtained from NOAA 1988 (NOAA'88) global radiosonde measurements. This NOAA'88 data set is widely used in radiation simulations and satellite remote sensing (e.g., Seemann et al., 2003) and covers both land and oceans. The data set has more than 5000 profiles, and about 1/3 of them are for cloudy skies. In cloudy cases, the NOAA'88 profiles can have up to two layers of clouds. Thus, the simulated results represent both clear and cloudy conditions. Since the model TOA (40km) height is much higher than that of radiosonde measurements, whenever there are no radiosonde upper atmospheric observations, interpolated climatological values of the upper atmosphere (McClatchey et al., 1972) are used. The weighting functions for the interpolation are decided from the surface air temperatures and pressures to meet the radiosonde measured weather conditions. In order to have large variations in surface air pressure, for each NOAA'88 measured profile, the surface pressure is randomly shifted by a Gaussian number with standard deviation 12mb, and the ratio of the shifted surface air pressure to the measured surface pressure is calculated. The atmospheric pressures in the measured profile above the surface are, then, adjusted to the values using the same ratio as that of the surface pressure.

For the analysis in this section, the radar system is assumed to fly on an aircraft at 15 km altitude with velocity 200 m/s, downward-looking and having a beamwidth of 3°, which produces a footprint of 785 m. The NOAA hurricane reconnaissance aircraft generally fly above 10 km height through and/or over hurricanes. Since this study is the first step in the model simulations for the radar system to show feasibility of the radar remote sensing for sea surface barometry, the 15 km altitude simulations provide us sufficient theoretical and technical insights for the radar sea surface pressure measurements. For other altitudes, the radar retrievals should have similar accuracy to those simulated here. During our simulation, since all wavelengths used in the radar system are very close to each other, we assume the surface reflection (or σ^0) to be the same (11 dB) for all frequency channels (Callahan et al., 1994). As we have showed in the previous section, the absolute magnitude of the surface reflectivity is not very important for surface pressure estimation as long as the spectrum dependence of σ^0 within the O_2 bands is negligible.

Simulated signals are analyzed in the form of relative received power (RRP), i.e., the ratio of the received and transmitted powers of the considered radar system. Since the system works at the O_2 absorption bands, the relative received powers are generally weak. Certain signal coding techniques for carrier frequencies, correlators for signal receiving and long-time (0.2s) averages of received powers are useful components for consideration for the radar system. Preliminary studies have disclosed advantages from a number of commonly employed radar techniques.

The radar-received signals reflected from sea surfaces, i.e. RRP values, used in this section are simulated through the complicated MWRT calculations discussed previously. With the RRP values, we calculate the radar differential absorption index, Ri, defined in equation 4. As shown above, the index and sea surface air pressure have a near-linear relationship, which points out the basic directions and sensitivities for surface air pressure remote sensing.

Atmospheric extinctions (or attenuations) vary dramatically at the O_2 band radar frequencies between 50.3 and 55.5 GHz. At the lowest frequency (50.3GHz), the atmospheric extinction optical depth is about 0.5, and at the highest frequency (55.5GHz), the optical depth goes sharply up to about 9. These two frequency cases represent the two extreme ends of weak and strong, respectively, atmospheric O_2 absorptions for our considered active microwave remote sensing of sea surface barometric pressure. With a weak O_2 absorption (i.e., small optical depth) radar signals would have significant influence from environments, such as atmospheric water vapor, cloud water amount and atmospheric temperature profile but transmitted powers used might be lower. While the atmospheric O_2 absorption is too strong, most of radar-transmitted powers would be close to attenuation, and small changes in surface air pressure (or column O_2 amount) would not produce significant differences in the received powers. This might be offset somewhat by using higher transmitted power. Thus at constant transmitter power levels, wavelengths with moderate to reasonably strong O_2 absorptions in the atmosphere are expected to serve our purpose best by giving a reasonable compromise between transmission and visibility.

Figure 8 shows examples of atmospheric extinction optical depths counted from TOA under clear conditions using the standard profiles (*McClatchey et al.* 1972). The three different color

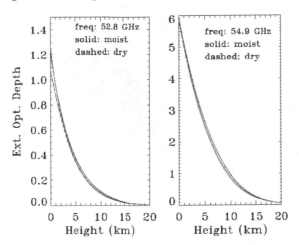

Fig. 8. Atmospheric extinction optical depths for various atmospheric temperatures and moisture levels at 52.8 and 54.9 GHz.

curves represent atmospheric surface temperatures of 280, 290 and 300K, respectively. It can be seen that these curves are very close each other, indicating atmospheric temperature effects are minimal. For channel 2 (i.e. 52.8GHz, left panel) cases, the optical depths for moist atmospheres (solid curves) with 40mm column water vapor are about 1.25 and only 0.1 higher than those of dry atmospheres. At 54.9GHz (right panel), the optical depths are increased considerably to about 6, and different temperature and moisture conditions have little effect on the total extinctions. For this frequency, the atmospheric extinctions of radar received signals due to double atmospheric path lengths reach about 50dB. This may require enhancements to the radar signals to control end to end noise, as mentioned before.

For tropical meteorological cases, such as hurricane cases, the changes in temperature and moisture profiles are even much smaller than those shown in the figure due to limited temperature and humidity conditions for the tropical storm development. To test accuracies of surface pressure measurements, a 15 dB SNR (signal-to-noise ratio) for radar-received signals is assumed for this primary study.

Figure 9 shows the simulated relationship between the differential absorption index (the logarithm of the radar return ratio of relative received powers at wavelengths 53.6 and 54.4 GHz and sea surface air pressure. Each point in the figure represents one adjusted NOAA'88 profile. As discussed above, good linear correlations of the two variables are further established by these simulations. A linear regression gives the root mean square (rms) error in sea surface air pressure estimates about 7.5 mb, which may be suitable for many meteorological uses. For frequencies of 53.6 and 54.9 GHz (Figure 10), simulated results (5.4 mb) are close to current theoretical O_2 A-band results. The best results (in Figure 11) we found are those from the differential absorption index 52.8 and 54.9GHz. The rms error in this case is about 4.1 mb. The tight linear relation between the sea surface air pressure and differential absorption index provides a great potential of remote sensing surface air pressure from airborne radar systems. Note that in Figs. 9-11, the dynamic range of sea surface barometric pressure is only from ~ 960mb to ~1050mb. The low end of the dynamic

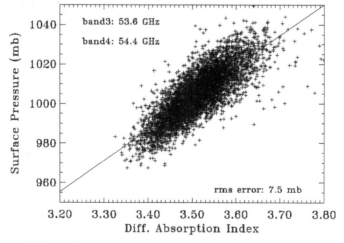

Fig. 9. Simulated relationship between the differential absorption index, the logarithm of the radar spectrum ratio at wavelengths 53.6 and 54.4 GHz , and surface air pressure.

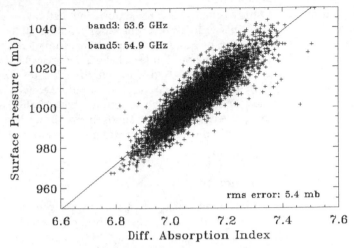

Fig. 10. Similar to Fig. 9, except frequencies are changed to 53.6 and 54.9 GHz.

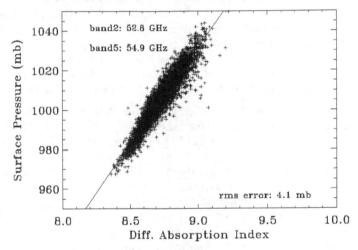

Fig. 11. Same as Fig. 10, except for 52.8 and 54.9GHz.

range of the sea surface pressure is significantly higher than some sea surface air pressures of hurricane centers. NOAA 1988 profiles were measured in generally average weather and meteorological environments, and were not taken from tropical storm cases. Thus, there were no extreme low sea surface air pressures in the NOAA data set. Actually, for tropical storm cases, the signal strength and SNR of the radar measurements at all O_2 band channels would be higher than those in normal conditions due to low atmospheric radar attenuation caused by low O_2 amounts (or the low hurricane center pressures). Also, the hurricane centers are generally clear skies. So, the accuracy of radar retrievals of the sea surface barometric pressure for hurricane center cases would be higher than those shown in the figures. The key to reach high accuracies of sea surface barometric pressure measurements is to have a high SNR of radar received powers reflected from sea surfaces.

This theoretical and modeling study establishes a remote sensing method for sea surface air pressure. Simulated results show that with an airborne radar working at about 53~55GHz O_2 absorption bands, the rms errors of the radar surface pressure estimations can be as small as 4~7mb. The considered radar systems should at least have 2 frequency channels to obtain the relative received power ratios of the two wavelengths. For the best simulated combination of 52.8 and 54.9 GHz channels, the power loss of radar received signals due to dual atmospheric path length absorptions could be as high as about 50 dB. High signal-to-noise ratios for radar reflected powers after these atmospheric absorptions will require modern radar technologies. In addition, careful radar design to insure stable instrument gain will be required.

4. DiBAR demonstration instrument

The goal in developing the demonstration instrument was to use commercial-of-the-shelf hardware wherever possible to develop the capability to collect differential absorption data that would verify the simulated differential absorption results, and to allow various measurement approaches to be assessed. An important operational characteristic for the radar, and determining factor in most design tradeoffs for the DiBAR system, is the SNR. The optimum channel to use in the O_2 absorption band from $50 \sim 56$ GHz is a function of the radar SNR, which depended on the surface reflectivity and the total atmospheric absorption. Thus, rather than selecting a set of frequencies bases on the microwave atmospheric absorption model, the demonstration instrument will have the flexibility to vary the measurement frequencies, and even to measure the differential absorption from 50 to 56 GHz and allow multiple processing and data analysis strategies to be evaluated for the same data set.

The basic instrument concept utilizes a Vector Network Analyzer (VNA) and a millimeter wave Up/Down Converter subsystem to enable operation from $50 \sim 56$ GHz. The millimeter wave Up/Down Converter will translate the VNA measurements to the O_2 absorption band, and provide very flexible signal processing options. As shown in Figure 12, the Up/Down Converter provides a millimeter power amplifier for the transmitter and a Low Noise Amplifier (LNA) for the receiver. The transmit power is selectable but the maximum is limited by the Q-band output amplifier to +14 dBm. The maximum transmit power and the receiver noise figure, 5.3 dB, will establish the SNR for our selected flight altitude. Our analysis indicates that for altitudes below 1000 m the SNR will be sufficient to verify the differential absorption across the O_2 absorption band. The transmit power can also be reduced during the flight to assess the impact of SNR on various data analysis approaches. Finally, to maximize isolation and eliminate the need for a Q-band transmit/receive (T/R) switch, the demonstration instrument transmitter and receiver are each fitted with an antenna.

The DiBAR demonstration instrument is extremely versatile and can be operated in several modes to emulate a wide range of radar modes and processing concepts. Several modes of operation can be used to collect absorption band data to increase probability of success and provide additional insight into the concept of differential absorption. The anticipated data sets will also provide insight into other phenomenon, at these frequencies, such as sea surface scattering. The instrument can be retrofitted with microwave switches to allow hardware gating, if required, to reduce any radar return other than the ocean surface. This option is not presently implemented.

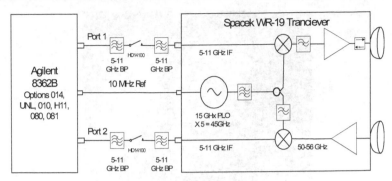

Fig. 12. DIBAR demonstration instrument block diagram.

For the data discussed here, the DiBAR instrument was operated in a stepped Continuous Wave (CW) mode using Fourier transform and windowing to produce software gating in the time domain. This processing minimized the effect of radar returns other than from the sea surface, or leakage between the transmitter and receiver.

4.1 Preliminary functional testing

Laboratory functional testing of the system such as, characterization of system linearity, noise figure, antenna gain, and isolation between antennas has been completed and reported elsewhere (Lawrence et al. 2007; Lin et al., 2006). Results of these tests were nominal with two minor exceptions. The frequency response of the Up/Down Converter, shown in figure 12, varied over the frequency range of 50 and 56 GHz by more that 12 dB. This change with frequency was larger that expected. However, it has been assumed that low altitude DiBAR data would be used to characterize the frequency response of the instrument during the flight tests. Therefore, as long as frequency response is stable, this should not affect the DiBAR demonstration flight tests. The leakage from the transmitter to the receiver within the Up/Down Converter enclosure was larger than mutual coupling between antennas. The impact of this leakage is minor. Our stepped CW measurement approach allowed software gating to suppress this term as long as the range to the target is more than about 10 to 15 m. Again, this had no impact on flight tests.

The assembled DiBAR demonstration radar is shown in figure 13 during a quick test using a water tower as a target to verify the operation of the radar. The DiBAR instrument collected 16001 stepped CW measurements for frequencies from 53 to 56 GHz. The Fourier transform of these data then results in a time domain representation of the radar return as a function of range. The resulting time domain data is shown in figure 14 and the large return from the water tower as well as the internal leakage term can clearly be seen in the figure.

The data in figure 14 may be helpful in illustrating the DiBAR measurement approach. The DiBAR instrument must provide precision measurements of the variation in the radar return as a function of frequency. Using a similar stepped CW measurement approach over the ocean, we can transform the data to the time domain, and then use windowing to minimize the effects of clutter. The windowed time domain data can then be transformed back to the frequency domain to measure the differential absorption index. An important assumption for our test flight planning is that the frequency response of the instrument will be

Fig. 13. DiBAR Demonstration Radar

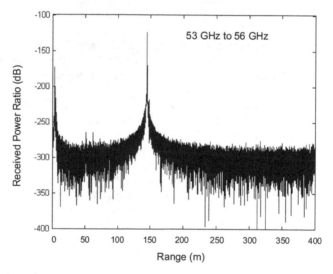

Fig. 14. Radar return from water tower vs. range

characterized by comparing stepped CW data at various flight altitudes. This of course assumes stability of the instrument frequency response. In order to verify the stability of the frequency response, the DiBAR instrument was moved into an anechoic chamber to

measure the backscatter from a conductive sphere in a stable and controlled environment. Unfortunately, the available chamber was not designed for millimeter wave frequencies, so precision radar cross section measurements or absolute calibration of the DiBAR instrument was not possible. However, while clutter was apparent in the radar measurements, the facility did provide a stable environment and was useful for the primary objective of characterizing the stability of the instrument.

The data was collected in the stepped CW mode using 16001 points from 50 to 56 GHz over several hours. The time domain result of a measurement of a 35.5 cm diameter sphere is shown in figure 15. The sphere can be seen at a range of approximately 22 m. The leakage term appears near zero range and the back wall of the facility is only a few meters further downrange than the sphere. Windowing was used to reduce the error due to these contaminating signals and the data is then transformed back to the frequency domain. Assuming the sphere is stationary, any change in the measured response can be attributed to variation in the end-to-end frequency response of the DiBAR demonstration instrument.

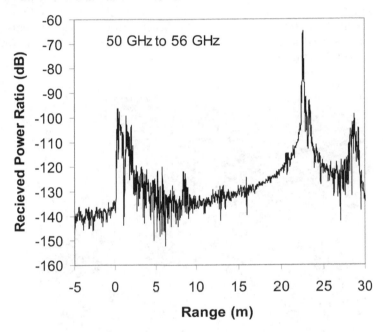

Fig. 15. Radar return from sphere vs. range

4.2 DIBAR flight test results

The initial flight-testing to verify the differential loss was accomplished utilizing a helicopter that provided several test flights over water in varying atmospheric and sea conditions. Several modifications to the DiBAR instrument were required for these tests. The integration of the DiBAR instrument on board the helicopter required the high gain antennas to be replaced with smaller horn antennas. The reduction in antenna gain results in reduced system dynamic range, and limits the maximum altitude where sufficient signal to noise ratio is available for useful pressure measurements. To minimize the impact of the antenna modification, the frequency sweep was increased from 53-56 GHz to 50-60 GHz for these flights. While the spectral response of the DiBAR instrument decreases above 56 GHz, the increased O_2 attenuation at these frequencies may be useful for the lower altitude operations. Analysis using an instrument model developed from laboratory testing and the microwave absorption model described in (Lin & Hu, 2005, Lawrence et al., 2007) suggests that this configuration of the instrument will provide an estimate of the differential O_2

absorption for an altitude of approximately 3000 feet (ft). Note that within US aviation industry aircraft altitude is reported in feet. Since this is the value recorded by the flight crew, altitude will be reported in feet in this description.

The demonstration DiBAR instrument was installed on a helicopter (Figure 16) for several test flights. Data was collected with in-situ estimated barometric pressure ranging from 1007 to 1028 mb. At each measurement site, the DiBAR instrument made three to five measurements of radar return for frequencies from 50 to 60 GHz. These measurements were performed while the helicopter was in a hover and each measurement set included altitudes from 500 to 5000 ft. These measurements were performed at each altitude with the helicopter at nominally the same location. The 500 ft altitude measurements for each measurement set was used to provide correction for sea surface reflectivity variations and spectral calibration of the instrument.

Fig. 16. DiBAR Instrument Installed in vehicle for initial flight tests.

The results for a data set performed on a day with an in-situ estimated barometric pressure at the measurement location of approximately 1018mb are shown in figure 17. DiBAR data for 2000, 3000, and 5000 ft altitudes is shown, as well as the modeled radar return. Three DiBAR measurements were performed at each altitude, and are indicated by the three different symbols in Figure 17. The predicted radar return (solid curve) is estimated using the radar equation for an extended target (sea surface) and the microwave absorption model adapted from (Lawrence et al., 2007; Lin & Hu, 2005). The measured transfer function of the DiBAR instrument was then combined with these models to estimate the expected radar return, shown in Figure 17 as the solid curve. The DiBAR measurements for each altitude are very repeatable, suggesting that the DiBAR instrument and the sea surface scattering characteristics were sufficiently stable. The reduced radar return as the measurement frequency increases can clearly be seen in Figure 17. This reduction is partially due to the increased O_2 attenuation discussed in section 3.

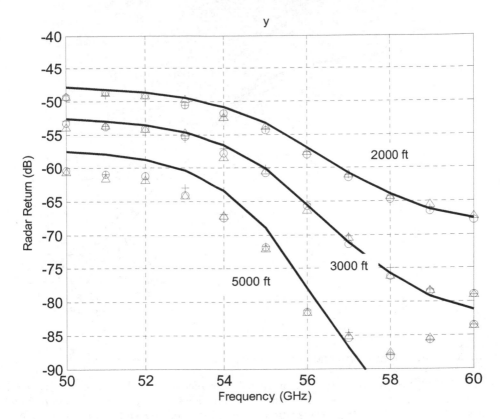

Fig. 17. Comparison of DiBAR measured return and model predictions

The measured results agree very well with the model for 2000 ft altitude measurements. The results for 3000 ft also agree well with the model for frequencies from 50 to 58 GHz. The difference between the measured and predicted values above 58 GHz is likely due to the noise floor of the modified DiBAR instrument. That is, due to the reduced antenna gain the signal to noise ratio of the DiBAR is insufficient at frequencies above 58 GHz at 3000 ft altitude and above 56 GHz at 5000 ft altitude. It appears that the optimum trade off between sufficient O_2 absorption (path length) and the noise floor of the DiBAR instrument for these flights occurs at an altitude of approximately 3000ft. Future flights with the high gain antennas will not have this limitation.

DiBAR data for 3000 ft from three difference days are shown in Figure 18. Three measurements are indicated for each day (symbols) as well as the predicted values (solid line). The increase in attenuation with increasing frequency can be seen in the data for all three days. Further, the attenuation appears to increase with increasing barometer pressure as would be expected. The difference between barometric pressures for each day is approximately 10 mb. While no statistical analysis was performed, the variation in the measured attenuation above 57 GHz appears to be on-the-order of the variation between each day. That is, the measurement-to-measurement variation was on the order of ± 5 mb

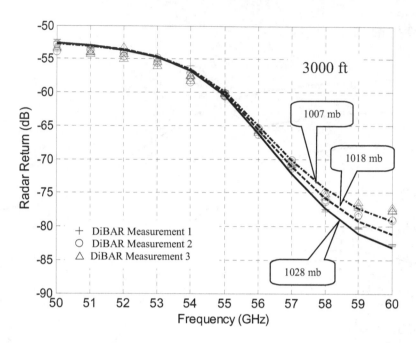

Fig. 18. Measured radar return and model predictions for three pressure days.

for the 3000 ft altitude data. The stability of these measurements over several minutes indicates that sea surface scattering can be assumed constant for these conditions. As discussed in Section 3 this increase in attenuation is expected to result in a linear change in differential absorption, Ri(f1,f2) defined in equation 4.

The differential absorption index is also provided by DiBAR measurements. The DiBAR demonstration instrument measures the radar return over the entire frequency band from 50 to 60 GHz. However, the differential attenuation index can be extracted from the data where the radar signals are sufficiently above the noise floor. For example, the differential absorption for f1= 53 GHz and f2=58 GHz, or Ri(53,58), can be found from Figure 18 by

subtracting the radar return for 58 GHz from that for 53 GHz. Figure 19 shows Ri(53,58) measured at altitudes of 1000, 2000, 3000, and 4000 ft. The measured data for the three pressure days are shown in the figure as well as the predicted Ri(53,58) using the instrument model and microwave atmospheric attenuation model discussed above. The figure illustrates the affect of increasing altitude. As the altitude increases the increased path length increases proportionality constant between Ri and P_o in equation (4). Thus, ignoring the receiver SNR, a less precise estimate of Ri is required for the same surface pressure precision at higher altitudes. Conversely, at 1000 ft larger changes in barometric pressure would be required to produce a detectable change in Ri. This demonstrates the impact of the reduction in antenna gain and limiting the useful measurement altitude to 3000 ft. However, the differential absorption index shown in Figure 19 agrees well with the predicted values for Ri, through 3000 ft altitude.

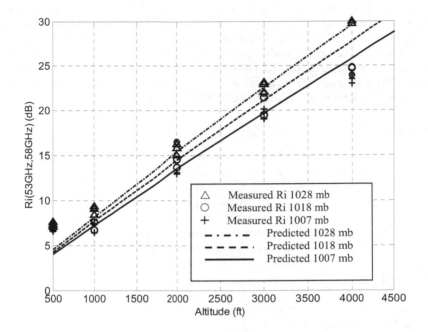

Fig. 19. DiBAR derived and predicted differential absorption coefficients.

5. Conclusions

The goal of the initial flight testing was to demonstrate differential radar measurement approach. The DiBAR measurements for the Chesapeake Bay at multiple altitudes demonstrated very good agreement between measured and predicted results for altitudes below approximately 3000 ft and for frequencies below 56 GHz. In addition, multiple measurements at these altitudes indicate little change over several minutes. This suggests that changes on the surface reflection coefficient over these time scales can be ignored for these surface conditions and spatial resolution. As expected, above 3000 ft the reduced antenna gain resulted in insufficient signal to noise ratio. However, the measured differential absorption index was in general agreement with the modeled values. Further, although beyond the scope of these initial flight tests, variations in the DiBAR measurements for 3000 ft measurements appear to be in the range ± 5 mb. These results are encouraging and consistent with our accuracy goal. Future flight testing should include an assessment of the barometric pressure measurement for high altitude and future satellite operations.

The initial flight testing described above successfully demonstrated the measurement approach. To fully demonstrate the measurement of surface level pressure will likely require flight data at altitudes between 5 kft and 15 kft using the original high gain antennas. An onboard calibration system should also be developed to eliminate the need for low altitude data to correct for changes to the spectral response of the instrument. In addition, while the existing demonstration DiBAR instrument is suitable to demonstrate the concept, a radar processor should be developed specifically for the differential absorption measurement to eliminate the need for the PNA. This would substantially reduce the weight and size of the instrument. This modification should not only eliminate the PNA, but should also be designed to enhance the stability of the instrument and enable the pulse operation to eliminate one of the antennas. While eventually funding will be required to develop an operational DiBAR instrument capable of operation at altitudes of 40 kft, these improvements may lead to moderate altitude flight opportunities.

6. References

Barton, I.J., and Scott, J.C. (1986). Remote measurement of surface pressure using absorption in the Oxygen A-band, *Appl. Opt.*, 25, 3502-3507.

Callahan, P.S., Morris, C.S. and Hsiao, S.V. (1994). Comparison of TOPEX/POSEIDON σ_0 and significant wave height distributions to Geosat, *J. Geophys. Res.*, 99, 25015-25024,.

Chou M-D (1990). Parameterization for the absorption of solar radiation by O2 and CO2 with application to climate studies. *J. Climate*, 3, 209-217.

Chou, M-D. (1992). A solar radiation model for climate studies. *J. Atmos. Sci.*, 49, 762-772.

Chou M-D and Suarez, M. J. (1994). *An efficient thermal infrared radiation parameterization for use in general circulation models*, NASA Tech Memo 104606.

Flower, D.A., and Peckham, G.E. (1978). *A microwave pressure sounder*, JPL Publication 78-68, CalTech, Pasadena, CA.

Ho, S.-P., Lin, B., Minnis, P., and Fan T.-F.(2003). Estimation of cloud vertical structure and water amount over tropical oceans using VIRS and TMI data, *J. Geophys. Res.*, 108 (D14), 4419, doi:10.1029/2002JD003298.

Huang, J., Minnis, P. , Lin, B., Yi, Y., Khaiyer, M.M., Arduini, R.F., Fan, A., Mace, G.G. (2005). Advanced retrievals of multilayered cloud properties using multi-spectral measurements, *J. Geophys. Res.*, 110, D15S18, doi:10.1029/2004JD005101.

Korb, C.L., and Weng, C.Y.(1982). A theoretical study of a two-wavelength lidar technique for the measurement of atmospheric temperature profiles, *J. Appl. Meteorol.*, 21, 1346-1355, 1982.

Lawrence, R., Fralick, D., Harrah, S., Lin, B., Hu, Y., Hunt, P., Differential absorption microwave radar measurements for remote sensing of atmospheric pressure, Proceedings of the IEEE International Geoscience and Remote Sensing Symposium, July 2007.

Liebe, H.(1989). MPM--An atmospheric millimeter-wave propagation model. *Int. J. Infrared and Millimeter Waves*, 10, 631-650, 1989.

Liebe, H., Hufford, G., and Manabe, T. (1991). A model for complex permittivity of water at frequencies below 1 THz, *Int. J. Infrared Millimeter Waves*, 12, 659-675.

Lin, B., and Rossow, W.B.(1994). Observations of cloud liquid water path over oceans: Optical and microwave remote sensing methods, *J. Geophys. Res.*, 99, 20907-20927.

Lin, B., and Rossow, W. B. (1996). Seasonal variation of liquid and ice water path in non-precipitating clouds over oceans, *J. Clim.*, 9, 2890-2902.

Lin, B., and Rossow, W. B. (1997). Precipitation water path and rainfall rate estimates over oceans using Special Sensor Microwave Imager and International Satellite Cloud Climatology Project data, *J. Geophys. Res.*, 102, 9359-9374.

Lin, B., Wielicki, B., Minnis, P., and Rossow, W.(1998a) Estimation of water cloud properties from satellite microwave, infrared and visible measurements in oceanic environments, 1. Microwave brightness temperature simulations, *J. Geophys. Res.*, 103, 3873-3886.

Lin, B., Minnis, P., Wielicki, B., Doelling, D. R., Palikonda, R., Young, D. F., and Uttal, T. (1998b) Estimation of water cloud properties from satellite microwave, infrared and visible measurements in oceanic environment, 2. Results, *J. Geophys. Res.*,103, 3887-3905.

Lin, B. and Minnis, P. (2000). Temporal variations of land surface microwave emissivities over the ARM southern great plains site, *J. App. Meteor.*, 39, 1103-1116.

Lin, B., Minnis, P., Fan, A., Curry, J., and Gerber, H. (2001). Comparison of cloud liquid water paths derived from in situ and microwave radiometer data taken during the SHEBA/FIREACE, Geophys. Res. Letter, 28, 975-978.

Lin, B. and Hu, Y. (2005). Numerical Simulations of Radar Surface Air Pressure Measurements at O_2 Bands, *IEEE Geosci. and Remote Sensing Letter*, 2, 324-328.

Lin, B., Harrah, S., Neece, R. Lawrence, R., and Fralick, D. (2006). *The Feasibility of Radar-Based Remote Sensing of Barometric Pressure, Final Report*, NASA Earth Science Technology Office, August 10, 2006.

McClatchey, R., Fenn, R., Selby, J., Voltz, E., and Garing, J. (1972). *Optical properties of the atmospheric*, Air Force Cambridge Research Laboratories Environmental Research Paper AFCRL-72-0497, No. 411, 108pp.

Rosenkranz, P. (1998). Water vapor microwave continuum absorption: A comparison of measurements and models, *Radio Sci.*, 33, 919-928.

Ray, P. (1972). Broadband complex refractive indices of ice and water, *Appl. Opt.*, 11, 1836-1844.

Seemann, S. W., Li, J., Menzel, W. P., and Gumley, L. E. (2003). Operational retrieval of atmospheric temperature, moisture, and ozone from MODIS infrared radiances, *J. Appl. Meteorol.*, 42(8), 1072-1091.

Singer, S.F. (1968). Measurement of atmospheric surface pressure with a satellite–borne laser, Appl. Opp. 7, 1125-1127.

Wang, D–H., Droegemeier, K. K., Jahn, D., Xu, K. –M., Xue, M., and Zhang, J. (2001). NIDS-based intermittent diabatic assimilation and application to storm-scale numerical weather prediction. 14th Conf. On Numerical Weather Prediction and 18th Conf. On Weather and Forecasting, Amer. Meteor. Soc., Ft. Lauderdale, FL, 2001.

Wang, D. –H., and Minnis, P. (2003). *4D Data Reanalysis/Assimilation with Satellite, Radar and the Extensive Field Measurements*, CRYSTAL-FACE Science Team Meeting, Salt Lake City, UT, 24-28 Feb. 2003.

Wu, M.-L. (1985). Remote sensing of cloud top pressure using reflected Solar radiation in the Oxygen A-band, *J. Clim. Appl. Meteor.*, 24, 539-546.

Xiao, Q., Zou, X., and Wang, B. (2000). Initialization and simulation of a landfalling hurricane using a variational bogus data assimilation scheme, *Monthly Weather Review*, 128, 2252-2269.

Xue, M., Wang, D. –H., Gao, J. –D., Brewster, K., and Droegemeier, K. K. (2003). The Advanced Regional Prediction System (ARPS): storm-scale numerical weather prediction and assimilation. *Meteor. Atmos. Physics*, 82, 139-170.

10

Three-Dimensional Lineament Visualization Using Fuzzy B-Spline Algorithm from Multispectral Satellite Data

Maged Marghany

Institute of Geospatial Science and Technology (INSTeG)
Universiti Teknologi Malaysia, UTM, Skudai, Johor Bahru
Malaysia

1. Introduction

A lineament is a linear feature in a landscape which is an expression of an underlying geological structure such as a fault. Typically a lineament will comprise a fault-aligned valley, a series of fault or fold-aligned hills, a straight coastline or indeed a combination of these features. Fracture zones, shear zones and igneous intrusions such as dykes can also give rise to lineaments. Lineaments are often apparent in geological or topographic maps and can appear obvious on aerial or satellite photographs. The term 'megalineament' has been used to describe such features on a continental scale. The trace of the San Andreas Fault might be considered an example. The Trans Brazilian Lineament and the Trans-Saharan Belt, taken together, form perhaps the longest coherent shear zone on the Earth, extending for about 4,000 km. Lineaments have also been identified on other planets and their moons. Their origins may be radically different from those of terrestrial lineaments due to the differing tectonic processes involved (Mostafa and Bishta, 2005; Semere and Ghebreab, 2006).

Accurate geological features mapping is critical task for oil exploration, groundwater storage and understanding the mechanisms of environmental disasters for instance, earthquake, flood and landslides. The major task of geologists is documentation of temporal and spatial variations in the distribution and abundance of geological features over wide scale. In this context, the major challenge is that most of conventional geological surveying techniques are not able to cover a wide region of such as desert in the Earth's surface. Quite clearly, to understand the mechanisms generations of geological features and their relationship with environmental disasters such as earthquake, landslide and flood, geological researchers must be able to conduct simultaneous measurements over broad areas of surface or subsurface of the Earth(Novak and Soulakellis 2000 and Marghany et al., 2009a).

This requires the collection of asset of reliable synoptic data that specify variations of critical geological environmental parameters over a wide region for discrete moments. In fact that geological features such as lineament and faults are key parameters that described the Earth generation or disaster mechanisms and significant indicator for oil explorations and groundwater storages (Semere and Ghebreab, 2006). Fortunately, the application of remote-sensing technology from space is providing geologists with means of acquiring these synoptic data sets.

1.1 Satellite remote sensing and image processing for lineament features detection

Lineaments are any linear features that can be picked out as lines (appearing as such or evident because of contrasts in terrain or ground cover on either side) in aerial or space imagery. If geological these are usually faults, joints, or boundaries between stratigraphic formations. Other causes of lineaments include roads and railroads, contrast-emphasized contacts between natural or man-made geographic features (e.g., fence lines), or vague "false alarms" caused by unknown (unspecified) factors. The human eye tends to single out both genuine and spurious linear features, so that some thought to be geological may, in fact, be of other origins (Semere and Ghebreab, 2006).

In the early days of Landsat, perhaps the most commonly cited use of space imagery in Geology was to detect linear features (the terms "linear" or "photolinear" are also used instead of lineaments, but 'linear' is almost a slang word) that appeared as tonal discontinuities. Almost anything that showed as a roughly straight line in an image was suspected to be geological. Most of these lineaments were attributed either to faults or to fracture systems that were controlled by joints (fractures without relative offsets) (Wang et al. 1990; Vassilas et al. 2002; Robinson et al., 2007).

Lineaments are well-known phenomena in the Earth's crust. Rocks exposed as surfaces or in road cuts or stream outcrops typically show innumerable fractures in different orientations, commonly spaced fractions of a meter to a few meters apart. These lineaments tend to disappear locally as individual structures, but fracture trends persist. The orientations are often systematic meaning, that in a region, joint planes may lie in spatial positions having several limited directions relative to north and to horizontal (Mostafa and Bishta, 2005). Where continuous subsurface fracture planes that extend over large distances and intersect the land surface produce linear traces (lineaments). A linear feature in general can show up in an aerial photo or a space images as discontinuity that is either darker (lighter in the image) in the middle and lighter (darker in the images) on both sides; or, is lighter on one side and darker on the other side. Obviously, some of these features are not geological. Instead, these could be fence lines between crop fields, roads, or variations in land use. Others may be geo-topographical, such as ridge crests, set off by shadowing. But those that are structural (joints and faults) are visible in several ways (Semere and Ghebreab, 2006; Zaineldeen 2011).

Lineament commonly are opened up and enlarged by erosion. Some may even become small valleys. Being zones of weak structure, they may be scoured out by glacial action and then filled by water to become elongated lakes (the Great Lakes are the prime example). Ground water may invade and gouge the fragmented rock or seep into the joints, causing periodic dampness that we can detect optically, thermally, or by radar. Vegetation can then develop in this moisture-rich soil, so that at certain times of year linear features are enhanced. We can detect all of these conditions in aerial or space imagery (Majumdar and Bhattacharya 1998; Katsuaki et al., 1995; Walsh 2000; Mostafa and Bishta, 2005; Semere and Ghebreab, 2006).

Consequently, optical remote sensing techniques over more than three decades have shown a great promise for mapping geological feature variations over wide scale (Mostafa and Bishta, 2005; Semere and Ghebreab, 2006; Marghany et al., 2009a). In referring to Katsuaki et al., (1995); Walsh (2000) lineament information extractions in satellite images can be divided broadly into three categories: (i) lineament enhancement and lineament extraction for characterization of geologic structure;(ii) image classification to perform geologic mapping or to locate spectrally anomalous zones attributable to mineralization (Mostafa et al., 1995; Süzen and Toprak 1998); and (iii) superposition of satellite images and multiple data such as geological, geochemical, and geophysical data in a geographical information system (Novak and Soulakellis 2000; Semere and Ghebreab 2006). Furthermore, remote sensing data assimilation in real time could be a bulk tool for geological features extraction and mapping. In this context, several investigations currently underway on the assimilation of both passive and active remotely sensed data into automatic detection of significant geological features i.e., lineament, curvilinear and fault.

Image processing tools have used for lineament feature detections are: (i) image enhancement techniques (Mah et al. 1995; Chang et al. 1998; Walsh 2000;Marghany et al., 2009b); and (ii) edge detection and segmentation (Wang et al. 1990; Vassilas et al. 2002; Mostafa and Bishta 2005). In practice, researchers have preferred to use the spatial domain filtering techniques in order to get ride of the artificial lineaments and to verify disjoint lineament pixels in satellite data (Süzen and Toprak 1998). Further, Leech et al., (2003) implemented the band-rationing, linear and Gaussian nonlinear stretching enhancement techniques to determine lineament populations. Won-In and Charusiri (2003) found that High Pass Filter enhancement technique provides accurate geological map. In fact, the High Pass filter selectively enhances the small scale features of an image (high frequency spatial components) while maintaining the larger-scale features (low frequency components) that constitute most of the information in the image.

Majumdar and Bhattacharya (1998) and Vassilas et al. (2002), respectively have used Haar and Hough transforms as edge detection algorithms for lineament detection in Landsat-TM satellite data. Majumdar and Bhattacharya (1998) reported that Haar transform is proper in extraction of subtle features with finer details from satellite data. Vassilas et al. (2002), however, reported that Hough transform is appropriate for fault feature mapping. Consequently, Laplacian, Sobel, and Canny are the major algorithms for lineament feature detections in remotely sensed data (Mostafa and Bishta 2005; Semere and Ghebreab, 2006; Marghany 2005).Recently Marghany and Mazlan (2010) proposed a new approach for automatic detection of lineament features from RADARSAT-1 SAR data. This approach is based on modification of Lee adaptive algorithm using convolution of Gaussian algorithm.

1.2 Problems for geological features extraction from remote sensing data

Geological studies are requiring standard methods and procedures to acquire precisely information. However, traditional methods might be difficult to use due to highly earth complex topography. Regarding the previous prospective, the advantage of satellite remote sensing in its application to geology is the wide coverage over the area of interest, where much accurate and useful information such as structural patterns and spectral features can be extracted from the imagery. Yet, abundance of geological features are not be fully understood. Lineaments are considered the bulk geological features which are still unclear in spite of they are useful for geological analysis in oil exploration. In this sense, the lineament extraction is very important for the application of remote sensing to geology. However the real meaning of lineament is still vague. Lineaments should be discriminated from other line features that are not due to geological structures. In this context, the lineament extraction should be carefully interpreted by geologists.

1.3 Hypothesis of study

Concerning with above prospective, we address the question of uncertainties impact on modelling Digital Elevation Model (DEM) for 3-D lineament visualization from multispectral satellite data without needing to include digital elevation data. This is demonstrated with LANDSAT-ETM satellite data using fuzzy B-spline algorithm (Marghany and Mazlan 2005 and Marghany et al., 2007). Three hypotheses are examined:

* lineaments can be reconstructed in Three Dimensional (3-D) visualization;
* Canny algorithm can be used as semiautomatic tool to discriminate between lineaments and surrounding geological features in optical remotely sensed satellite data; and
* uncertainties of DEM model can be solved using Fuzzy B-spline algorithm to map spatial lineament variations in 3-D.

2. Study area

The study area is located in Sharjah Emirates about 70 Km fromSharjah city. It is considered in the alluvium plain for central area of UAE and covers an Area of 1800 Km² (60 km x 30 km) within boundaries of latitudes 24° 12'N to 24°.23'N and longitudes of 55°.51'E to 55° 59' E (Fig. 1). The northern part of UAE is formed of the Oman mountains and the marginal hills extends from the base of the mountains and (alluvium plain) to the south western sand dunes (Figs 2 and 3) such features can be seen clearly in Wadi Bani Awf, Western Hajar (Fig.2). Land geomorphology is consisted of structural form, fluvial, and Aeolian forms(sand dunes). According to Maged et al., (2009) structural form is broad of the Oman mountains and JabalFayah (Fig.4) which are folded structure due collusion of oceanic crust and Arabian plate (continental plate). Furthermore, the mountain is raised higher than 400 m above sea level and exhibit parallel ridges and high–tilted beds. Many valleys are cut down the mountains, forming narrow clefts and there are also intermittent basins caused by differential erosion. In addition, the Valley bases are formed small caves. Stream channels have been diverted to the southwest and they deposited silt in the tongue -shaped which lies between the dunes. Further, Aeolian forms are extended westwards from the Bahada plain, where liner dunes run towards the southwest direction in parallel branching pattern (Fig. 3) with relative heights of 50 meters. Nevertheless, the heights are decreased towards the southeast due to a decrease in sand supply and erosion caused by water occasionally flowing from the Oman mountains. Moreover, some of the linear dunes are quite complex due to the development of rows of star dunes along the top of their axes. Additionally, inter dunes areas are covered by fluvial material which are laid down in the playas formed at the margins of the Bahadas plain near the coastline. The dunes changes their forms to low flats of marine origin and their components are also dominated by bioclastics and quartz sands (Marghany and Mazlan 2010).

Fig. 1. Location of Study area.

Fig. 2. Geologic fault feature along Oman mountain.

Fig. 3. Dune forms on Oman mountain base.

Fig. 4. Sand dune feature along Jabal Fayah.

3. Data sets

In study, there are two sort of data have been used. First is satellite data which is involved LANDSAT Enhanced Thematic Mapper (ETM) image with pixel resolution of 30 m which is acquired on 14:07, 18 December 2004 (Fig.5). It covers area of 24° 23' N, 55° 52' E to 24° 17' N and 55° 59' E (Fig.5). Landsat sensors have a moderate spatial-resolution. It is in a polar, sun-synchronous orbit, meaning it scans across the entire earth's surface. With an altitude of 705 kilometres +/- 5 kilometres, it takes 232 orbits, or 16 days, to do so. The satellite weighs 1973 kg, is 4.04 m long, and 2.74 m in diameter. Unlike its predecessors, Landsat 7 has a solid state memory of 378 gigabits (roughly 100 images). The main instrument on board Landsat 7 is the Enhanced Thematic Mapper Plus (ETM+).

The main features of LANDSAT-7 (Robinson et al., 2007) are

- A panchromatic band with 15 m (49 ft) spatial resolution (band 8).
- Visible (reflected light) bands in the spectrum of blue, green, red, near-infrared (NIR), and mid-infrared (MIR) with 30 m (98 ft) spatial resolution (bands 1-5, 7).
- A thermal infrared channel with 60 m spatial resolution (band 6).
- Full aperture, 5% absolute radiometric calibration.

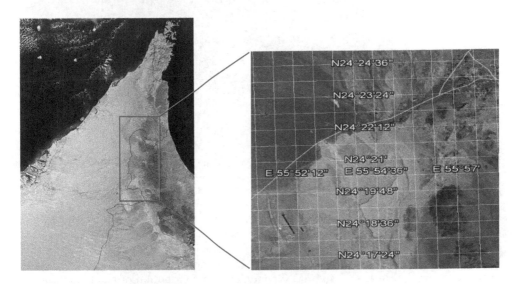

Fig. 5. LANDSAT satellite data used in this study

Second is ancillary data which are contained digital topographic, geological maps, well logs and finally ground water data. Furthermore, ancillary data such as topography map of scale 1: 122,293 used to generate Digital Elevation Model (DEM) of selected area. Bands 1,2,3,5 and 7 are selected to achieve the objective of this study. According to Marghany et al., (2009) these bands can provide accurate geological information. Finally, the Digital Elevation Model (DEM) is acquired from SRTM data (Fig.6).

4. Model for 3-D lineament visualization

The procedures have been used to extract lineaments and drainage pattern from LANDSAT ETM satellite image were involved image enhancement contrast, stretching and linear enhancement which were applied to acquire an excellent visualization. In addition, automatic detection algorithm Canny are performed to acquire excellent accuracy of lineament extraction (Mostafa et al., 1995). Two procedures have involved to extract lineaments from LANDSAT ETM data. First is automatic detection by using automatic edge detection algorithm of Canny algorithm. Prior to implementations of automatic edge detection processing, LANDSAT ETM data are enhanced and then geometrically corrected. Second is implementing fuzzy B-spline was adopted from Marghany et al., (2010) to reconstruct 3D geologic mapping visualization from LANDSAT ETM satellite data.

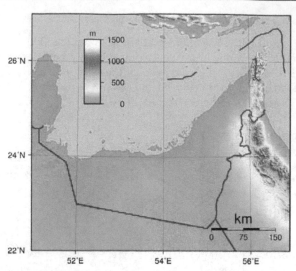

Fig. 6. Topographic map of United Arab Emirates that created with GMT from SRTM data

4.1 Histogram equalization

Following Marghany et al., (2009) histogram equalization is applied to LANDSAT TM image to obtain high quality image visualization. An image histogram is an analytic tool used to measure the amplitude distribution of pixels within an image. For example, a histogram can be used to provide a count of the number of pixels at amplitude 0, the number at amplitude 1, and so on. By analyzing the distribution of pixel amplitudes, you can gain some information about the visual appearance of an image. A high-contrast image contains a wide distribution of pixel counts covering the entire amplitude range. A low contrast image has most of the pixel amplitudes congregated in a relatively narrow range (Süzen et al., 1998 and Gonzalez and Woods 1992).

4.2 Canny algorithm

According to Canny (1986),the Canny edge detector uses a filter based on the first derivative of a Gaussian, because it is susceptible to noise present on raw unprocessed image data, so to begin with, the raw image is convolved with a Gaussian filter. The result is a slightly blurred version of the original which is not affected by a single noisy pixel to any significant degree. According to Deriche (1987) the edge detection operator (Roberts, Prewitt, Sobel for example) returns a value for the first derivative in the horizontal direction (G_y) and the vertical direction (G_x). From this the edge gradient and direction (Θ) can be determined:

$$|G| = \sqrt{G_x{}^2 + G_y{}^2} \tag{1}$$

In fact, equation 1 is used to estimate the gradient magnitude (edge strength) at each point can be found to find the edge strength by taking the gradient of the image. Typically, an approximate magnitude is computed using

$$|G| = |G_x| + |G_y| \tag{2}$$

Equation 2 is faster to be computed.

$$\theta = \arctan\left(\frac{G_y}{G_x}\right) \tag{3}$$

The direction of the edge θ is computed using the gradient in the G_x and G_y directions. However, an error will be generated when sum X is equal to zero. So in the code, there has to be a restriction set whenever this takes place. Whenever the gradient (G) in the x direction is equal to zero, the edge direction has to be equal to 90 degrees or 0 degrees, depending on what the value of the gradient in the y-direction is equal to. If G_y has a value of zero, the edge direction will equal 0 degrees. Otherwise the edge direction will equal 90 degrees (Deriche 1987).

According to Gonzalez and Woods (1992),three criteria are used to improve edge detection. The first and most obvious is low error rate. It is important that edges occurring in images should not be missed and that there be NO responses to non-edges. The second criterion is that the edge points be well localized. In other words, the distance between the edge pixels as found by the detector and the actual edge is to be at a minimum. A third criterion is to have only one response to a single edge. This was implemented because the first 2 were not substantial enough to completely eliminate the possibility of multiple responses to an edge (Canny 1986).

4.3 The fuzzy B-splines algorithm

The fuzzy B-splines (FBS) are introduced allowing fuzzy numbers instead of intervals in the definition of the B-splines. Typically, in computer graphics, two objective quality definitions for fuzzy B-splines are used: triangle-based criteria and edge-based criteria (Marghany et al., 2009). A fuzzy number is defined using interval analysis. There are two basic notions that we combine together: confidence interval and presumption level. A confidence interval is a real values interval which provides the sharpest enclosing range for current gradient values.

An assumption μ -level is an estimated truth value in the [0, 1] interval on our knowledge level of the topography elevation gradients (Anile 1997). The 0 value corresponds to minimum knowledge of topography elevation gradients, and 1 to the maximum topography elevation gradients. A fuzzy number is then prearranged in the confidence interval set, each one related to an assumption level $\mu \in [0, 1]$. Moreover, the following must hold for each pair of confidence intervals which define a number: $\mu \succ \mu' \Rightarrow d \succ d'$.

Let us consider a function $f : d \to d'$, of N fuzzy variables $d_1, d_2, ..., d_n$. Where d_n are the global minimum and maximum values topography elevation gradients along the space. Based on the spatial variation of the topography elevation gradients, the fuzzy B-spline algorithm is used to compute the function f (Marghany et al., 2010). Follow Marghany et al., (2010) $d(i,j)$ is the topography elevation value at location i,j in the region D where i is the horizontal and j is the vertical coordinates of a grid of m times n rectangular cells. Let N be the set of eight neighbouring cells. The input variables of the fuzzy are the amplitude differences of water depth d defined by (Anile et al. 1997):

$$\Delta d_N = d_i - d_0, N = 1, ..., 4 \tag{4}$$

where the d_i, $N=1$, 4 values are the neighbouring cells of the actually processed cell d_0 along the horizontal coordinate i. To estimate the fuzzy number of topography elevation d_j which is located along the vertical coordinate j, we estimated the membership function values μ and μ' of the fuzzy variables d_i and d_j, respectively by the following equations were described by Rövid et al. (2004)

$$\mu = \max\left\{\min\left\{m_{pl}(\Delta d_i) : d_i \in N_i\right\}; N = 1...., 4\right\} \tag{5}$$

$$\mu' = \max\left\{\min\left\{m_{LNl}(\Delta d_i) : d_i \in N_i\right\}; N = 1...., 4\right\} \tag{6}$$

Equations 5 and 6 represent topography elevation in 2-D, in order to reconstruct fuzzy values of topography elevation in 3-D, then fuzzy number of digital elevation in z coordinate is estimated by the following equation proposed by Russo (1998) and Marghany et al., (2010),

$$d_z = \Delta\mu MAX\{m_{LA}\left|d_{i-1,j} - d_{i,j}\right|, m_{LA}\left|d_{i,j-1} - d_{i,j}\right|\} \tag{7}$$

where d_z fuzzy set of digital elevation values in z coordinate which is function of i and j coordinates i.e. $d_z = F(d_i, d_j)$. Fuzzy number F_O for water depth in i,j and z coordinates then can be given by

$$F_O = \{\min(d_{z_0}, ..., d_{z_\Omega}), \max(d_{z_0}, ..., d_{z_\Omega})\} \tag{8}$$

where $\Omega = 1, 2, 3, 4,$

The fuzzy number of water depth F_O then is defined by B-spline in order to reconstruct 3-D of digital elevation. In doing so, B-spline functions including the knot positions, and fuzzy set of control points are constructed. The requirements for B-spline surface are set of control points, set of weights and three sets of knot vectors and are parameterized in the p and q directions.

Following Marghany et al., (2009b) and Marghany et al., (2010), a fuzzy number is defined whose range is given by the minimum and maximum values of digital elevation along each kernel window size. Furthermore, the identification of a fuzzy number is acquired to summarize the estimated digital elevation data in a cell and it is characterized by a suitable membership function. The choice of the most appropriate membership is based on triangular numbers which are identified by minimum, maximum, and mean values of digital elevation estimated. Furthermore, the membership support is the range of digital elevation data in the cell and whose vertex is the median value of digital elevation data (Anile et al. 1997).

5. Three-dimensional lineament visualization

5.1 3-D lineament visulization using classical method

Fig. 4 shows the Digital Elevation Model is derived from SRTM data that covered area of approximately 11 km². Clearly, DEM varies between 319-929 m and maximum elevation value of 929 m is found in northeast direction of UAE. Therefore, SRTM has promised to produce DEM with root mean square error of 16 m (Nikolakopoulos et al., 2006). In addition, Oman mountain is dominated by highest DEM value of 929 m which is shown parallel to coastal zone of Arabian Gulf. The DEM is dominated by spatial variation of the topography features such as ridges, sand dunes and steep slopes. As the steep slopes are clearly seen within DEM of 400 m (Fig,7). According to Zaineldeen (2011), the rocks are well bedded massive limestones with some replacement chert band sand nodules. The limestone has been locally dolomitized.

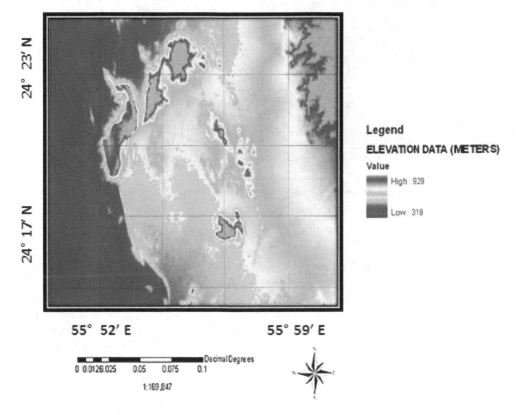

Fig. 7. DEM for study area.

Fig. 8 shows the supervised classification map of LANDSAT ETM satellite data. It clear that the vegetation covers are located in highest elevation as compiled with Fig. 7 while highlands are located in lowest elevation with DEM value of 660 m. The supervised

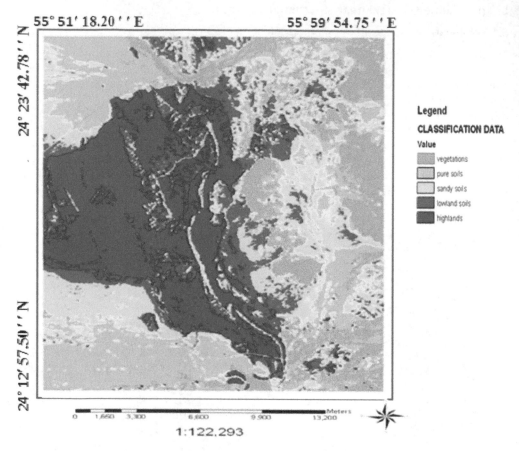

Fig. 8. Supervised map results.

classification shows a great fault moves through a highland area. According to Robinson et al., (2007) , TM bands 7 (2.08–2.35 mm),4 (0.76–0.90 mm),and 2(0.50–0.60 mm)are appropriate for geological features detection because they have low-correlation and produce high-contrast. In this regard, band 2 is useful for rock discrimination, band 4 for land/water contrasts, and band 7 for discrimination of mineral and rock types. Further, TM bands 7 are also able to imagined crest dunes parallel with tens kilometres of length. This feature is clear in northern part of Fig.8 and located in high land of DEM of 900 m. This finding confirm the study of Robinson et al., (2007).

Fig. 9 shows the output result mapping of lineaments using composite of bands 3,4 , 5 and 7 in LANDSAT TM satellite data. The appearance of lineaments in LANDSAT TM satellite image are clearly distinguished. In addition, area adjacent to the mountainous from Manamh (northward), Flili village in the (southward) has high density of lineaments due to the westward compressive force between the oceanic crust and Arabian plate, such as fractures and faults and drainage pattern that running in the buried fault plains (filled weathered materials coming from Oman mountains) (Fig. 9). The lineaments are associated with fractures and faults which are located in northern part of Fig. 9. In fact that Canny algorithm first is smoothed the image to eliminate and noise. It then finds the image gradient to highlight regions with high spatial derivatives. The algorithm then tracks along these regions and suppresses any pixel that is not at the maximum (non-maximum suppression). The gradient array is further reduced by hysteresis. According to Deriche (1987), hysteresis is used to track along the remaining pixels that have not been suppressed. Hysteresis uses two thresholds and if the magnitude is below the first threshold, it is set to zero (made a non-edge).

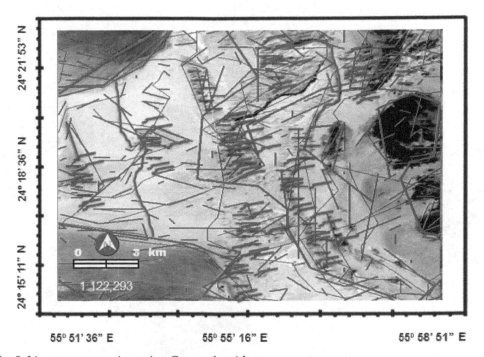

Fig. 9. Lineament mapping using Canny algorithm.

Further, If the magnitude is above the high threshold, it is made an edge. And if the magnitude is between the 2 thresholds, then it is set to zero unless there is a path from this pixel to a pixel with a gradient above threshold. In order to implement the canny edge detector algorithm, a series of steps must be followed. The first step is to filter out any noise in the original image before trying to locate and detect any edges. In fact, the Gaussian filter can be computed using a simple mask, it is used exclusively in the Canny algorithm. Once a suitable mask has been calculated, the Gaussian smoothing can be performed using standard convolution methods. According to Marghany et al., (2009), LANDSAT TM data can be used to map geological features such as lineaments and faults. This could be contributed to that composite of bands 3,4,5 able and 7 in LANDSAT TM satellite data are appropriate for mapping of geologic structures (Katsuaki and Ohmi 1995; Novak and Soulakellis 2000; Marghany et al., 2009). Consequently, the ground resolution cell size of LANDSAT TM data is about 30 m. This confirms the study of Robinson et al., (2007).

Fig. 10 shows the lineament distribution with 3D map reconstruction using SRTM and LANDSAT TM bands 3,4,5, and 7. It is clear that the 3D visualization discriminates between different geological features. It can be noticed the faults, lineament and infrastructures clearly (Figure 10b). This study agrees with Marghany et al., (2009). It can be confirmed that the lineament are associated with faults and it also obvious that heavy capacity of lineament occurrences within the Oman mountain. This type of lineament can be named as mountain lineament.

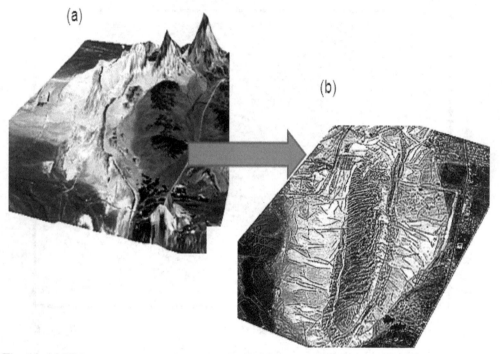

Fig. 10. (a) 3D image reconstruction using SRTM data and (b) lineament distribution over 3D image.

According to Robinson et al., (2007) and Marghany et al., (2009) the mountain is raised higher than 400 m above sea level and exhibit parallel ridges and high-tilted beds. Many valleys are cut down the mountains, forming narrow clefts and small caves. The fluvial forms are consisted of streams channels which are flowed from Oman mountains have and spread out into several braided channels at the base of the mountains from the Bahada and Playa plains (Figure 11). Stream channels have been diverted to the southwest and they deposited silt in the tongue -shaped which lies between the dunes.

Further, Aeolian forms are extended westwards from the Bahada plain, where liner dunes run towards the southwest direction in parallel branching pattern (Fig. 11) with relative heights of 50 meters. Nevertheless, the heights are decreased towards the southeast due to a

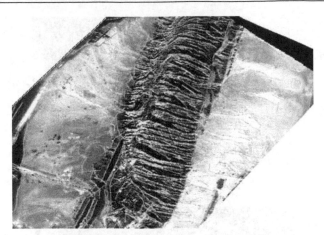

Fig. 11. 3D image and lineament distribution from Canny algorithm.

decrease in sand supply and erosion caused by water occasionally flowing from the Oman mountains. Moreover, some of the linear dunes are quite complex due to the development of rows of star dunes along the top of their axes. Additionally, inter dunes areas are covered by fluvial material which are laid down in the playas formed at the margins of the Bahadas plain near the coastline. The dunes changes their forms to low flats of marine origin and their components are also dominated by bioclastics and quartz sands (Marghany et al., 2009 and Zaineldeen 2011).

5.2 3-D lineament visulization using fuzzy B-spline technique

Fig. 12 shows the result acquires by using fuzzy B-spline algorithm. It is clear that the 3D visualization discriminates between different geological features. It can be noticed the faults, lineament and infrastructures clearly (Fig. 12c). This is due to the fact that the fuzzy B-splines considered as deterministic algorithms which are described here optimize a triangulation only locally between two different points (Fuchs et al., 1977; Anile et al., 1995;Anile, 1997; Marghany et al., 2010; Marghany and Mazlan 2011). This corresponds to the feature of deterministic strategies of finding only sub-optimal solutions usually. The visualization of geological feature is sharp with the LANDSAT TM satellite image due to the fact that each operation on a fuzzy number becomes a sequence of corresponding operations on the respective μ-levels and the multiple occurrences of the same fuzzy parameters evaluated as a result of the function on fuzzy variables Keppel 1975; Anile et al., 1995; Magrghany and Mazlan 2011).

It is very easy to distinguish between smooth and jagged features. Typically, in computer graphics, two objective quality definitions for fuzzy B-splines were used: triangle-based criteria and edge-based criteria. Triangle-based criteria follow the rule of maximization or minimization, respectively, of the angles of each triangle (Fuchs et al., 1977). The so-called max-min angle criterion prefers short triangles with obtuse angles. This finding confirms those of Keppel 1975 and Anile 1997. Table 1 confirms the accurate of fuzzy B-spline to eliminate uncertainties of 3-D visualization. Consequently, the fuzzy B-spline shows higher performance with standard error of mean of 0.12 and bias of 0.23 than SRTM technique. In

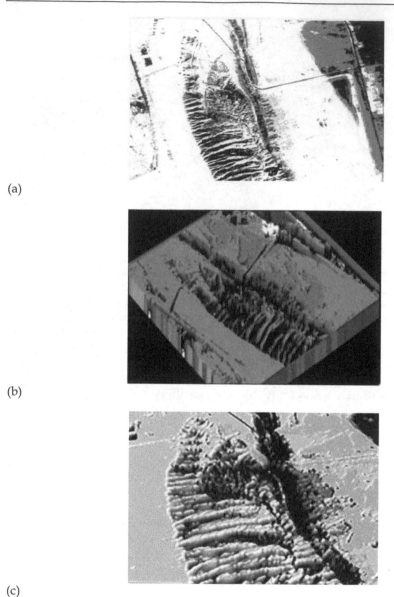

(a)

(b)

(c)

Fig. 12. (a): LANDSAT ETM satellite data and (b): 3D fuzzy B-spline visualization and (c): Zoom area of lineaments and fault

fact, Fuzzy B-splines provide both a continuous approximating model of the experimental data and a possibilistic description of the uncertainty in such DEM. Approximation with FBS provides a fast way to obtain qualitatively reliable descriptions whenever the introduction of a precise probabilistic DEM is too costly or impossible. In this study, fuzzy B-spline algorithm produced 3-D lineament visulization without need to ground geological survey. In fact fuzzy B-spline algorithm is able to keep track of uncertainty and provide tool for representing spatially clustered geological features. This advantage of fuzzy B-spline is not provided in Canny algorithm and DEM produced by SRTM data.

Statistical Parameters	3-D Visualization	
	Fuzzy B-spline	SRTM
Bias	0.23	0.63
Standard error of the mean	0.12	0.56

Table 1. Statistical Comparison of 3-D computer visualization using Fuzzy-B-spline and SRTM.

6. Conclusions

This study has demonstrated a method to map lineament distributions in United Arab Emirates (UAE) using LANDSAT-TM satellite data. In doing so, 3D image reconstruction is produced using SRTM data. Then Canny algorithm is implemented for lineament automatic detection from LANDSAT TM bands of 3,4,5,and 7. The results show that the maximum DEM value of 929 m is found in the northeast direction of UAE. The vegetation covers are dominated feature in the highest DEM while highlands are located in lowest elevation of 660 m. In addition, Canny algorithm has detected automatically lineament and fracture features. Therefore, 3D visualization is discriminated between lineament and fault features. The results show that the highest spatial distribution of lineaments are appeared in Oman mountain which are named by lineament mountain. In conclusion, the integration between Digital Elevation Model (DEM) and Canny algorithm can be used as geomatic tool for lineament automatic detection in 3D visualization. Further, a fuzzy B-spline algorithm is used to reconstruct Three Dimensional (3D) visualization of geologic feature spatial variations with standard error of mean of 0.12 and bias of 0.23. In conclusion, combination between Canny algorithm and DEM generated by using fuzzy B-spline could be used as an excellent tool for geologic mapping.

7. References

Anile, A. M, (1997). *Report on the activity of the fuzzy soft computing group*, Technical Report of the Dept. of Mathematics, University of Catania, March 1997, 10 pages.

Anile, AM, Deodato, S, Privitera, G, (1995) *Implementing fuzzy arithmetic*, Fuzzy Sets and Systems, 72,123-156.

Anile, A.M., Gallo, G., Perfilieva, I., (1997). *Determination of Membership Function for Cluster of Geographical data*. Genova, Italy: Institute for Applied Mathematics, National Research Council, University of Catania, Italy, October 1997, 25p., Technical Report No.26/97.

Canny, J., A, (1986). Computational Approach To Edge Detection. IEEE Transactions on Pattern Analysis and Machine Intelligence. PAMI-8 (6), pp. 679-698.

Chang, Y.,Song, G., Hsu, S., (1998). Automatic Extraction of Ridge and Valley Axes Using the Profile Recognition and Polygon-Breaking Algorithm. *Computers and Geosciences*. 24, (1), pp. 83-93.

Deriche, R., (1987). Using Canny's criteria to derive a recursively implemented optimal edge detector. *International Journal of Computer Vision*. 1 (2), pp. 167-187.

Forster, B.C., (1985). Mapping Potential of Future Spaceborne Remote Sensing System. Procs. of 27th Australia Survey Congress, Alice Springs, Institution of Surveyors, Australia, Australia, 109-117.

Fuchs, H. Z.M. Kedem, and Uselton, S.P., (1977). Optimal Surface Reconstruction from Planar Contours. *Communications of the ACM*, 20, 693-702.

Gonzalez, R., and R. Woods (1992).Digital Image Processing, 3rd edition, Addison-Wesley Publishing Company. pp:200-229.

Guenther, G.C., Cunningham, A.G., LaRocque, P. E., and Reid, D. J. (2000). Proceedings of EARSeL-SIG-Workshop LIDAR,Dresden/FRG,EARSeL , Strasbourg, France,June 16 – 17, 2000.

Keppel, E. (1975). Approximation Complex Surfaces by Triangulations of Contour Lines. *IBM Journal of Research Development*, 19, pp: 2-11.

Katsuaki, K., N., Shuichi, and M., Ohmi ,(1995). Lineament analysis of satellite images using a segment tracing algorithm (STA). *Computers and Geosciences*.Vol. 21, No. 9, pp. 1091-I 104.

Leech, D.P., Treloar, P.J., Lucas, N.S., Grocott, J., (2003). Landsat TM analysis of fracture patterns: a case study from the Coastal Cordillera of northern Chile. *International Journal of Remote Sensing*, 24 (19),pp.3709-3726.

Marghany, M., (2005).Fuzzy B-spline and Volterra algorithms for modelling surface current and ocean bathymetry from polarised TOPSAR data. *Asian Journal of Information Technology*. 4, pp: 1-6.

Marghany M., and Hashim, M.,(2006). Three-dimensional reconstruction of bathymetry using C-band TOPSAR data. Photogrammetrie Fernerkundung Geoinformation. pp: 469-480.

Marghay, M., M., Hashim and Crackenal, A., (2007). 3D Bathymetry Reconstruction from AIRBORNE TOPSAR Polarized Data. In: Gervasi, O and Gavrilova, M (Eds.): Lecture Notes in Computer Science. Computational Science and Its Applications – ICCSA 2007, ICCSA 2007, LNCS 4705, Part I, Volume 4707/2007, Springer-Verlag Berlin Heidelberg, pp. 410–420, 2007.

Marghany, M. S., Mansor and Hashim, M., (2009a). Geologic mapping of United Arab Emirates using multispectral remotely sensed data. *American J. of Engineering and Applied Sciences*. 2, pp: 476-480.

Marghany,M., M. Hashim and Cracknell A (2009b). 3D Reconstruction of Coastal Bathymetry from AIRSAR/POLSAR data. *Chinese Journal of Oceanology and Limnology*.Vol. 27(1), pp.117-123.

Marghany, M. and M. Hashim (2010). Lineament mapping using multispectral remote sensing satellite data. *International Journal of the Physical Sciences* Vol. 5(10), pp. 1501-1507.

Marghany, M., M. Hashim and Cracknell A. (2010). 3-D visualizations of coastal bathymetry by utilization of airborne TOPSAR polarized data. *International Journal of Digital Earth*, 3(2):187 – 206.

Mah, A., Taylor, G.R., Lennox, P. and Balia, L., (1995). Lineament Analysis of Landsat Thematic Mapper Images, Northern Territory, Australia. *Photogrammetric Engineering and Remote Sensing*, 61(6),pp. 761-773.

Majumdar, T.J., Bhattacharya, B.B., (1988). Application of the Haar transform For extraction of linear and anomalous over part of Cambay Basin, India. *International Journal of Remote Sensing*. 9(12),pp. 1937-1942.

Mostafa, M.E. and M.Y.H.T. Qari, (1995).An exact technique of counting lineaments. *Engineering Geology*. 39 (1-2), pp. 5-15.

Mostafa, M.E. and A.Z. Bishta, (2005). Significant of lineament pattern in rock unit classification and designation: A pilot study on the gharib-dara area. Northen eastern Desert, Egypt. *International Journal of Remote Sensing*. 26 (7), pp. 1463 – 1475.

Novak, I.D. and N. Soulakellis, (2000). Identifying geomorphic features using Landsat-5/TM data processing techniques on lesvos, Greece. *Geomorphology*. 34: 101-109.

Nikolakopoulos, K. G.; Kamaratakis, E. K; Chrysoulakis, N. (2006). "SRTM vs ASTER elevation products. Comparison for two regions in Crete, Greece". *International Journal of Remote Sensing*. 27 (21), 4819–4838.

Semere, S. and W. Ghebreab, (2006). Lineament characterization and their tectonic significance using Landsat TM data and field studies in the central highlands of Eritrea. *Journal of African Earth Sciences*. 46 (4), pp. 371-378.

Süzen, M.L. and V. Toprak, (1998).Filtering of satellite images in geological lineament analyses: An application to a fault zone in central Turkey. *International Journal of Remote Sensing*. 19 (6), pp. 1101-1114.

Russo, F., (1998).Recent advances in fuzzy techniques for image enhancement. IEEE Transactions on Instrumentation and measurement. 47, pp: 1428-1434.

Robinson, C.A. F.El-Baz, T.M.Kuskyb, M.Mainguet, F.Dumayc, Z.AlSuleimani, A.Al Marjebye (2007). Role of fluvial and structural processes in the formation of the Wahiba Sands, Oman: A remote Sensing Prospective. *Journal of Arid Environments*. 69,676–694.

Rövid, A., Várkonyi, A.R. andVárlaki, P., (2004). 3D Model estimation from multiple images," IEEE International Conference on Fuzzy Systems, FUZZ-IEEE'2004, July 25-29, 2004, Budapest, Hungary, pp. 1661-1666.

Vassilas, N., Perantonis, S., Charou, E., Tsenoglou T., Stefouli, M., Varoufakis, S., (2002). Delineation of Lineaments from Satellite Data Based on Efficient Neural Network and Pattern Recognition Techniques. *2nd Hellenic Conf. on AI, SETN-2002*, 11-12 April 2002, Thessaloniki, Greece, Proceedings, Companion Volume,355-366.

Walsh, G.J. and S.F. Clark Jr., (2000). Contrasting methods of fracture trend characterization in crystalline metamorphic and igneous rocks of the Windham quadrangle, New Hampshire. Northeast. *Northeastern Geology and Environmental Sciences*. 22 (2), pp. 109-120.

Won-In, K., Charusiri, P., (2003). Enhancement of thematic mapper satellite images for geological mapping of the Cho Dien area, Northern Vietnam. *International Journal of Applied Earth Observation and Geoinformation*, Vol. 15, 1-11.

Zaineldeen U. (2011) Paleostress reconstructions of Jabal Hafit structures,Southeast of AlAin City, United Arab Emirates (UAE). *Journal of African Earth Sciences*. 59,323–335

Permissions

The contributors of this book come from diverse backgrounds, making this book a truly international effort. This book will bring forth new frontiers with its revolutionizing research information and detailed analysis of the nascent developments around the world.

We would like to thank Boris Escalante-Ramírez, for lending his expertise to make the book truly unique. He has played a crucial role in the development of this book. Without his invaluable contribution this book wouldn't have been possible. He has made vital efforts to compile up to date information on the varied aspects of this subject to make this book a valuable addition to the collection of many professionals and students.

This book was conceptualized with the vision of imparting up-to-date information and advanced data in this field. To ensure the same, a matchless editorial board was set up. Every individual on the board went through rigorous rounds of assessment to prove their worth. After which they invested a large part of their time researching and compiling the most relevant data for our readers. Conferences and sessions were held from time to time between the editorial board and the contributing authors to present the data in the most comprehensible form. The editorial team has worked tirelessly to provide valuable and valid information to help people across the globe.

Every chapter published in this book has been scrutinized by our experts. Their significance has been extensively debated. The topics covered herein carry significant findings which will fuel the growth of the discipline. They may even be implemented as practical applications or may be referred to as a beginning point for another development. Chapters in this book were first published by InTech; hereby published with permission under the Creative Commons Attribution License or equivalent.

The editorial board has been involved in producing this book since its inception. They have spent rigorous hours researching and exploring the diverse topics which have resulted in the successful publishing of this book. They have passed on their knowledge of decades through this book. To expedite this challenging task, the publisher supported the team at every step. A small team of assistant editors was also appointed to further simplify the editing procedure and attain best results for the readers.

Our editorial team has been hand-picked from every corner of the world. Their multi-ethnicity adds dynamic inputs to the discussions which result in innovative outcomes. These outcomes are then further discussed with the researchers and contributors who give their valuable feedback and opinion regarding the same. The feedback is then collaborated with the researches and they are edited in a comprehensive manner to aid the understanding of the subject.

Apart from the editorial board, the designing team has also invested a significant amount of their time in understanding the subject and creating the most relevant covers. They scrutinized every image to scout for the most suitable representation of the subject and create an appropriate cover for the book.

The publishing team has been involved in this book since its early stages. They were actively engaged in every process, be it collecting the data, connecting with the contributors or procuring relevant information. The team has been an ardent support to the editorial, designing and production team. Their endless efforts to recruit the best for this project, has resulted in the accomplishment of this book. They are a veteran in the field of academics and their pool of knowledge is as vast as their experience in printing. Their expertise and guidance has proved useful at every step. Their uncompromising quality standards have made this book an exceptional effort. Their encouragement from time to time has been an inspiration for everyone.

The publisher and the editorial board hope that this book will prove to be a valuable piece of knowledge for researchers, students, practitioners and scholars across the globe.

List of Contributors

Josué Álvarez-Borrego
CICESE, División de Física Aplicada, Departamento de Óptica, México

Beatriz Martín-Atienza
Facultad de Ciencias Marinas, UABC, México

Hooman Latifi
Dept. of Remote Sensing and Landscape Information Systems, University of Freiburg, Germany

Vladimir Lukin, Nikolay Ponomarenko and Dmitriy Fevralev
National Aerospace University, Ukraine

Kacem Chehdi and Benoit Vozel
University of Rennes 1, France

Andriy Kurekin
Plymouth Marine Laboratory, UK

Anna Brook and Marijke Vandewal
Royal Military Academy, CISS Department, Brussels, Belgium

Eyal Ben-Dor
Remote Sensing Laboratory, Department of Geography and Environment, Tel-Aviv University, Tel-Aviv, Israel

Rolando D. Navarro, Jr., Joselito C. Magadia and Enrico C. Paringit
University of the Philippines, Diliman, Quezon City, Philippines

Pau Bergada, Rosa Ma Alsina-Pages, Carles Vilella and Joan Ramon Regué
La Salle - Universitat Ramon Llull, Spain

Christian Rogaß, Daniel Spengler, Mathias Bochow, Karl Segl, Sigrid Roessner, Robert Behling, Hans-Ulrich Wetzel, Katia Urata and Hermann Kaufmann
Helmholtz Centre Potsdam, GFZ German Research Centre for Geosciences, Germany

Angela Lausch and Daniel Doktor
Helmholtz Centre for Environmental Research, UFZ Germany

Andreas Hueni
Remote Sensing Laboratories, University of Zurich, Switzerland

Ken C. K. Lee
Department of Computer and Information Science, University of Massachusetts Dartmouth, North Dartmouth, USA

Mao Ye and Wang-Chien Lee
Department of Computer Science and Engineering, The Pennsylvania State University, University Park, USA

Roland Lawrence
Old Dominion University, USA

Bin Lin and Steve Harrah
NASA Langley Research Center, USA

Qilong Min
SUNY at Albany, USA

Maged Marghany
Institute of Geospatial Science and Technology (INSTeG), Universiti Teknologi Malaysia, UTM, Skudai, Johor Bahru, Malaysia